Research Topics in Wind Energy

Volume 2

Series editor

Joachim Peinke, University of Oldenburg, Oldenburg, Germany
e-mail: peinke@uni-oldenburg.de

For further volumes:
http://www.springer.com/series/11859

About this Series

The series Research Topics in Wind Energy publishes new developments and advances in the fields of Wind Energy Research and Technology, rapidly and informally but with a high quality. Wind Energy is a new emerging research field characterized by a high degree of interdisciplinarity. The intent is to cover all the technical contents, applications, and multidisciplinary aspects of Wind Energy, embedded in the fields of Mechanical and Electrical Engineering, Physics, Turbulence, Energy Technology, Control, Meteorology and Long-Term Wind Forecasts, Wind Turbine Technology, System Integration and Energy Economics, as well as the methodologies behind them. Within the scope of the series are monographs, lecture notes, selected contributions from specialized conferences and workshops, as well as selected PhD theses. Of particular value to both the contributors and the readership are the short publication timeframe and the worldwide distribution, which enable both wide and rapid dissemination of research output. The series is promoted under the auspices of the European Academy of Wind Energy.

Michael Hölling · Joachim Peinke
Stefan Ivanell
Editors

Wind Energy – Impact of Turbulence

Springer

Editors
Michael Hölling
Institute of Physics and ForWind
University of Oldenburg
Oldenburg
Germany

Joachim Peinke
Institute of Physics and ForWind
University of Oldenburg
Oldenburg
Germany

Stefan Ivanell
Gotland University
Visby
Sweden

ISSN 2196-7806 ISSN 2196-7814 (electronic)
ISBN 978-3-642-54695-2 ISBN 978-3-642-54696-9 (eBook)
DOI 10.1007/978-3-642-54696-9
Springer Heidelberg New York Dordrecht London

Library of Congress Control Number: 2014933402

Printed on acid-free paper

Springer is part of Springer Science+Business Media (www.springer.com)

Preface

The seminar on "Wind Energy and the Impact of Turbulence on the Conversion Process" took place in Oldenburg, Germany, spring 2012. It was one of the first scientific meetings devoted to the common topic of wind energy and basic turbulence. The established community of researchers working on the challenging puzzle of turbulence for decades met the quite young community of researchers, who face the upcoming challenges in the fast growing field of wind energy applications. In the last few years it became more and more evident that the efficient conversion of wind energy by wind turbines has to face the challenging questions of turbulence. Although wind turbines become bigger and bigger, now having diameters close to 200m, they are always operating close to the ground in windy and gusty regions, respectively. From the fluid mechanical point of view, wind turbines are large machines operating in the fully turbulent atmospheric boundary layer. In particular they are facing small-scale turbulent inflow conditions. It is one of the central puzzle in basic turbulence research to achieve a fundamental understanding of the peculiarities of small-scale turbulence. Besides the inflow condition the resulting aerodynamics around the wind turbine's blades and the forces transmitted into the machinery need to be better understood in this context of puzzling inflow conditions. This is a big challenge due to the multi-scale properties of the incoming wind field ranging from local flow conditions on the profile up to the interaction of wake flows in wind farms.

This conference attracted high interest. It was supported from three societies, namely the EUROMech, EAWE and ECOFTATC. As organizers we were quite surprised that more than 100 participants from about 15 countries came to this conference. The intensive discussions of experts were a special mark of this conference as well as the active participation, which lasted until the last lecture. Thus by this conference the actuality of this new field became obvious.

Michael Hölling
Joachim Peinke
Stefan Ivanell

Contents

Part I: Rotor Aerodynamics

Part II: Turbulence

Part I

Rotor Aerodynamics

Fatigue Damage Mitigation by the Integration of Active and Passive Load Control Techniques on Wind Turbines

C.L. Bottasso, F. Campagnolo, A. Croce, and C. Tibaldi

Abstract. The synergistic use of passive (by bend-twist coupling) and active (by individual blade pitch control) load mitigation is considered in this work. It is found that while both contribute to load reduction, they have opposite effects on the actuator duty cycle, and this induces an additional unexpected pay-off from the combination of these two technologies.

1 Introduction

Load reduction techniques for wind turbines can be broadly categorized into two families: active and passive. The former aims at mitigating loads by actively controlling the machine (for example by changing the blade pitch and the generator torque, or by moving flaps and/or tabs or other active devices), while the latter is based on the idea of designing a structure that, when aerodynamically loaded, deforms so as to induce a load reduction.

Active load control systems exploit a power source to move the whole blade, or a part of it, so as to reduce loads due to turbulence, gusts and asymmetry in the inflow. Generally these systems require sensors to measure the wind turbine response and drive, through a feedback loop, a suitable motion strategy. Recently, several individual pitch control (IPC) formulations have been proposed, demonstrating a significant potential for fatigue load reduction [1, 2, 11, 7, 8, 9]. In fact, changing the pitch of each blade independently allows for the reduction of the lowest load harmonics, including their mean value. However, one of the possible drawbacks of

C.L. Bottasso · F. Campagnolo · A. Croce
Politecnico di Milano, Milan, Italy
e-mail: carlo.bottasso@polimi.it

C. Tibaldi
Risœ-DTU, Roskilde, Denmark
e-mail: tibl@risoe.dtu.dk

M. Hölling et al. (eds.), *Wind Energy – Impact of Turbulence*,
Research Topics in Wind Energy 2,
DOI: 10.1007/978-3-642-54696-9_1,

active load control is its effect on the actuators and their design: constantly pitching the blade increases the actuator duty cycle (ADC), and hence its wear. This, in turn, increases the cost of actuators, inducing a tradeoff with the advantages brought by IPC.

On the other hand, passive load alleviation by bend-twist coupling (BTC), obtained by exploiting the anisotropic mechanical properties of composite materials [12, 10, 6, 5] is very attractive for wind energy applications: in fact there are no actuators which may fail, no moving parts which may wear out, and no need for sensors, all characteristics that are very interesting whenever simplicity, low maintenance and high availability are key to reducing the cost of energy. Furthermore, BTC blades not only experience reduces loads but also require less ADC, since the blades self-react to turbulent fluctuations in the wind.

In this paper we explore the possible existing synergies between BTC and IPC. In fact, BTC and IPC can both mitigate loads, but BTC reduces ADC while IPC increases it. The combination of the two technologies can lead not only to improved load reduction performance, but also avoids excessive actuator activity thanks to the passive load control built into the blade design.

2 Approach and Methods

Blade design is here performed with a constrained optimization-based procedure [3, 4]. All design requirements are treated as constraints, therefore all converged solutions are viable according to the conditions that have been imposed by the designer. The code performs the design using a multi-level approach. The method includes 2D finite element models for sectional characterization and analysis, aero-servo-elastic multibody models for load calculation according to certification rules, and detailed 3D finite element models for detailed stress-strain, fatigue and buckling analysis. The optimizer iterates among the various levels, so as to deliver a cost-minimizing design solution that also satisfies all desired design constraints at the finest description level, i.e. the detailed 3D model.

The individual pitch controller is based on the architecture proposed in Ref. [1]. The controller is driven by blade root moments, which are Coleman transformed into a nacelle-fixed frame of reference to yield the rotor tilt and yaw moments. By adjusting the IPC gains, one can tune within a certain range the level of individual blade pitch activity, consequently affecting the level of load reduction and of ADC increase. For this study, two different sets of gains were selected: the first has lower gains and therefore only a moderate ADC increase, while the second one has higher gains and a more aggressive behavior.

3 Design of Coupled Blades for Passive Load Mitigation

The baseline uncoupled design used for this study is that of a 45 meter rotor blade for a Class-IIIA 2MW HAWT. All coupled blades are based on the same design load cases, constraints (5m maximum tip deflection, first flap frequency placement, max stress/strain and fatigue) and material properties of the uncoupled one.

At first, we consider the case where fibers are rotated in the skin and/or the spar caps for the whole span-wise extension of the blade according to six distinct configurations. Figure 1 at left reports the percent mass increase of the six optimal designs with respect to the uncoupled baseline model. It appears that mass increases rapidly with increasing fiber rotation angle. For example, this effect is evident by looking at the trend for the three solutions with fiber rotations in the skin (Sk+10, Sk+20 and Sk+30), and the trend of the two solutions with fiber rotations in the spar caps (SC+05 and SC+10). It also appears that the mass rate of increase is much higher for spar cap fiber rotations than for skin ones. These effects are due to the fact that this blade is mainly driven by frequency placement and maximum tip deflection constraints, so that, even if envelope loads and fatigue are reduced due to BTC, the blade structural components have to be thicker to restore the sectional stiffness and satisfy the design constraints. Therefore, since bending stiffness is largely dictated by the spar caps, the use of large fiber angles in the spar caps should be avoided to limit the impact on the blade mass. The figure also shows that by distributing the coupling between skin and spar caps, as done in the sixth design configuration Sk+20&SC+05, allows one to keep the angle in the spar caps close to small values, which in turn helps in limiting the reduction of sectional bending stiffness.

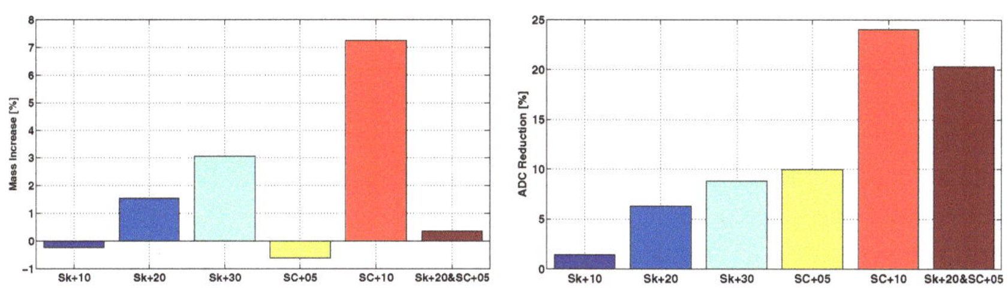

Fig. 1 Percent blade mass increase (left) and ADC percent reduction (right) with respect to the uncoupled baseline model

A very interesting aspect of BTC blades is their positive effects on ADC. In fact, Fig. 1 at right shows that for all coupled models there is a large reduction in the duty cycle of the pitch actuator. This is due to the passive control effect of coupled blades: since these adaptive blades self-respond to turbulent fluctuations in the wind, the active control system pitches the blades less (since it feels less the effects of turbulence through its feedback loop), thereby reducing the effort on the pitch system.

The effects of BTC on the hub envelope loads are illustrated in the left part of Fig. 2, while the right part reports the DEL reductions at the tower root with respect to the baseline configuration. Other loads are not shown here for brevity, but in all cases one can notice a mitigation of the loads due to the effects of bend-twist coupling.

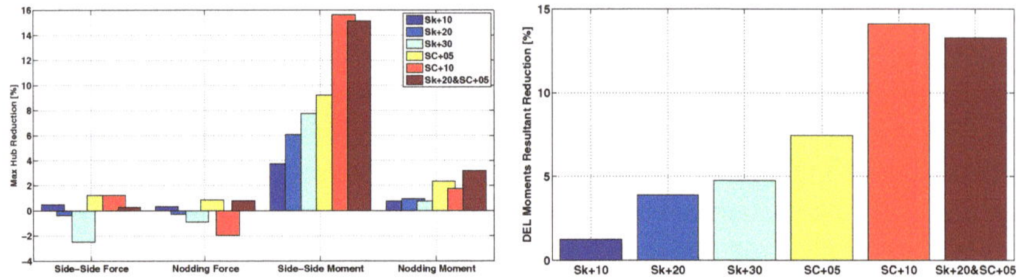

Fig. 2 Envelope load reductions at the hub (left) and DEL reductions at the tower root (right)

It was found that rotating the fibers away from the pitch axis in the region close to the root does not really contribute much to load reduction, while it may affect the frequency placement, maximum tip deflection and fatigue constraints. Indeed, these are the main reasons why the previously studied BTC blades weigh more than the uncoupled one. To correct for this problem, we modified blade model Sk+20&SC+05, chosen because it exhibits good quality metric improvements with only a modest mass increase, by using fiber rotations starting only outboard from a given span-wise location. Even in this case, each blade was obtained with the previously discussed constrained optimization procedure, and therefore each represents a constraint satisfying optimal design.

The results show that partial coupling has a beneficial effect on blade weight. Specifically, the minimum blade weight is achieved for rotations starting at 30% span from the root; this configuration exhibits a mass decrease of about 2.8% with respect to the uncoupled baseline and a decrease of about 3.2% with respect to the fully coupled solution. Looking in general at the loads on the various components of the machine (not shown here for brevity), one may notice that load reduction diminishes with the fiber rotation starting section moving outward. This is because the resulting blades are less coupled, and therefore there is less passive load reduction. Considering ADC, here again fiber rotation sections close to the root only suffer a modest loss of performance with respect to the fully coupled blade, while the more outboard sections significantly affect coupling, and hence incur in significant performance penalties.

4 Integration of Passive and Active Load Control

In the following, six combinations of blade and controller designs are considered. The baseline is taken to be the uncoupled blade with a collective pitch controller. In addition, we consider the partially coupled blade with the collective trimmer, termed BTC in the following figures. The uncoupled blade used with a first set of less aggressive gains is labeled IPC1, and IPC2 when the more aggressive set of gains is used. Similarly, the partially coupled blade is termed BTC+IPC1 when used with the first set of gains, and BTC+IPC2 with the second.

All configurations achieve essentially the same power production, with differences in their respective AEPs below 1%. However, when used with the uncoupled baseline blade, the less aggressive tuning of IPC1 generates ADC increases in the full power region in excess of 50%, while the more aggressive tuning of IPC2 brings the ADC increase to over 130%. Such large increases obtained with a standard blade would need to be accounted for during the design of the pitch system, increasing its wear and cost. On the other hand, the lifetime ADC for the BTC blade used with collective pitch control shows a reduction of almost 20%, while for BTC+IPC2 (more aggressive case) it increases of around 25% and for BTC+IPC1 (less aggressive case) it has only a slight increase of around 5% with respect to the uncoupled blade with collective pitch. This last result shows that the combination of the passive and active control strategies leads to an actuator duty cycle that is essentially equal to the one of the baseline blade with purely collective blade pitch control.

Figure 3 shows, for the six combinations, the hub DELs and mean moments vs. mean wind speed. It appears that, here again, the active and passive technologies have synergistic effects. In fact, models BTC+IPC1 and BTC+IPC2 have a higher load reduction than models BTC, IPC1 and IPC2, i.e. when using only either a passive or an active solution. Furthermore, it should be stressed once again that the load reductions achieved by BTC+IPC1 are obtained with roughly the same ADC as the baseline case.

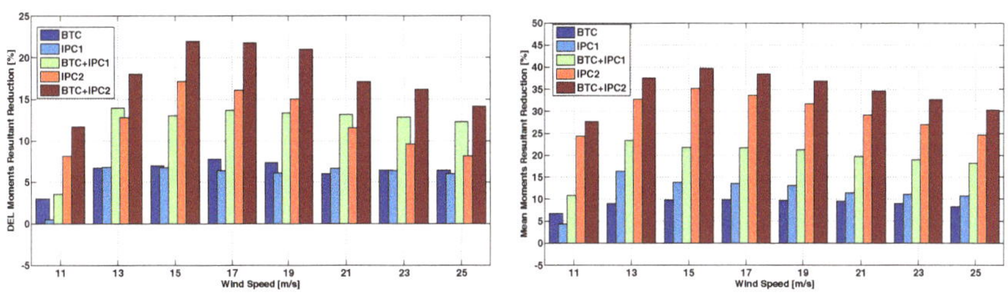

Fig. 3 Hub DELs (left) and mean moments (right) reductions with respect to the baseline model vs. mean wind speed

5 Conclusions

Based on the results of the present study, it is possible to draw the following conclusions:

- The synergistic use of fiber rotations in the skin and spar caps is beneficial in terms of blade weight. In fact, fiber rotations in the skin allow one to limit rotations in the spar caps, which otherwise would need to be thickened so as to contrast bending stiffness loss.
- Avoiding fiber rotations from the blade root to about the span of maximum chord is also beneficial in terms of blade weight. In fact the additional couplings generated in this region by fiber rotations are not only modest, but are also more than

offset by increased fatigue damage and reduced bending stiffness, which both lead to weight increases.
- BTC blades not only reduce maximum and fatigue loads, but also imply a reduced pitch ADC. In fact, the blade self-reacts to wind fluctuations, this way taking some of the burden off the controller. Since IPC tends to increase ADC, which is one of the main reasons why IPC has not yet been widely adopted by manufacturers, one might argue that BTC blades might also be used to make IPC more economically viable in future designs.
- The combination of passive, via BTC, and active, via IPC, technologies leads to load reductions that are higher than it would be possible by the use of each single one of the two approaches.

References

1. Bossanyi, E.: Individual blade pitch control for load reduction. Wind Energy 6, 119–128 (2003)
2. Bossanyi, E.: Further load reductions with individual pitch control. Wind Energy 8, 481–485 (2005)
3. Bottasso, C.L., Campagnolo, F., Croce, A.: Multi-disciplinary constraint optimization of wind turbines. Multibody System Dynamics (2011), doi:10.1007/s11044-011-9271-x
4. Bottasso, C.L., Campagnolo, F., Croce, A., Dilli, S., Nielsen, M.B.: Optimization of Wind Turbine Rotor Blades by Integrated Sectional/Multibody/3DFEM Analysis. DIA-SR 12-01. Department of Aerospace Engineering, Politecnico di Milano (2012)
5. Bottasso, C.L., Campagnolo, F., Croce, A., Tibaldi, C.: Optimization-based study of bend-twist coupled rotor blades for passive and integrated passive/active load alleviationy. Wind Energy (2012) (accepted) (to appear)
6. Capellaro, M., Kühn, M.: Boundaries of bend twist coupling. In: The Science of Making Torque from Wind, Crete, Greece (2010)
7. Geyler, M., Caselitz, P.: Individual blade pitch control design for load reduction on large wind turbines. In: EWEC 2007, Milano, Italy (2007)
8. Kanev, S.: van Engelen. T.: Exploring the limits in individual pitch control. In: EWEC 2009, Marseille, France (2009)
9. Leithead, W., Neilson, V., Dominguez, S.: Alleviation of unbalanced rotor loads by single blade controllers. In: EWEC 2009, Marseille, France (2009)
10. Lobitz, D.W., Veers, P.S., Eisler, G.R., Laino, D.J., Migliore, P.G., Bir, G.: The Use of Twist-Coupled Blades to Enhance the Performance of Horizontal Axis Wind Turbines. Sandia Report SAND 2001-1303. Sandia National Laboratories, Albuquerque, NM (2001)
11. van Engelen, T.: Design model and load reduction assessment for multi-rotational model individual pitch control. In: EWEC 2006, Athens, Greece (2006)
12. Veers, P.S., Bir, G., Lobitz, D.W.: Aeroelastic tailoring in wind-turbine blade application. In: Windpower 1998, AWEA, Bakersfield, CA (1998)

DES Study of Airfoil Lift Coefficient Sensitivity to Flow Turbulence

Elia Daniele, Iván Herráez, Bernhard Stoevesandt, and Joachim Peinke

Abstract. The aerodynamic behavior of wind turbines is strongly influenced by the turbulence level. However, the design of the rotor blades is usually based on experimental results of airfoils operating under laminar conditions. This leads to great uncertainties in the design process, which in turn make wind turbines less reliable and cost-effective. In this work a DES numerical study of the flow around a Wortmann FX 79-W-151A airfoil is performed for different turbulence intensities. Special attention is paid to the resulting loads. The simulations are then compared and validated with already available load measurements. The aim of this work is on one hand to gain a better understanding of the aerodynamics of an airfoil working in a turbulent flow. On the other hand, it is also of great interest to see up to which degree the numerical simulations are able to predict the force coefficients.

Elia Daniele
Department of Aerospace Engineering, University of Naples "Federico II",
Via Claudio 21 - 80125 Naples, Italy
e-mail: elia.daniele@unina.it

Iván Herráez
ForWind Center for Wind Energy Research, Ammerlnder Heerstr. 136 - 26129
Oldenburg, Germany
e-mail: ivan.herraez@forwind.de

Bernhard Stoevesandt
Fluid and System Dynamics Department, Fraunhofer IWES,
Carl-von-Ossietzky-Str. 9/11 - 26129 Oldenburg, Germany
e-mail: bernhard.stoevesandt@iwes.fraunhofer.de

Joachim Peinke
Physics Department, Carl von Ossietzky University, Carl-von-Ossietzky-Str. 9/11 - 26129
Oldenburg, Germany
e-mail: peinke@uni-oldenburg.de

M. Hölling et al. (eds.), *Wind Energy – Impact of Turbulence,*
Research Topics in Wind Energy 2,
DOI: 10.1007/978-3-642-54696-9_2, © Springer-Verlag Berlin Heidelberg 2014

1 Introduction

Blade designers usually have to rely on airfoil characteristics measured in laminar flow. This is a great drawback in the case of wind turbines, due to the fact that they typically operate under turbulent conditions. The great uncertainties derived from this mismatching between available experimental data and required turbulent data, affect negatively the aerodynamic design of wind turbines and consequently its performance and cost-effectiviness. Performing new experiments is very costly and time-consuming. Therefore, numerical models are often used for estimating the required airfoil characteristics.

Most commonly, RANS simulations are chosen as a good compromise between accuracy and computational cost, even if the prediction capabilities of this kind of simulation for separated flows are known to be very limited. LES simulations offer a good accuracy even at post-stall angles of attack but their computational cost is too high for real engineering applications [2]. A very promising alternative is a hybrid RANS-LES approach known as DES (Detached Eddy Simulation). This kind of simulation was first introduced by Spalart [14] in the year 1997 and was improved in the year 2006 by the same author [16]. Johansen [5] et alii and Li [6] came to the conclusion in their studies of NACA profiles, that especially in the post-stall region the use of DES makes a big difference. Bertagnolio et alii [1] described the grid requirements to be considered for performing successful DES simulations on airfoils. Gilling et alii [3] performed remarkable in-depth studies of airfoils in turbulent inflow.

Previous experimental work for airfoil characterization has also mainly been performed without studying the influence of turbulence [19, 12, 20, 8].

In opposition to most of the mentioned works, where turbulence intensity was not a key point of interest, our goal is to asses the sensitivity of the airfoil characteristics to the flow turbulence by means of DES simulations. The simulations, which are run with the numerical toolbox OpenFOAM [10], rely on the Spalart-Allmaras one-equation model [13]. In order to estimate their reliability a validation with available measurements on a Wortmann FX 79-W-151A airfoil [11] is carried out.

2 Numerical Methods

The chosen hybrid RANS-LES method, called Detached-Eddy Simulation (DES), has been firstly introduced by Spalart [14], and it is intended to act as a sub-grid scale model in regions where the grid density fulfills the LES requirements, while it acts as a simple RANS model in regions where the grid is too coarse. Moreover, a more rigorous use of the DES method requires that the turbulent boundary layer is *modelled* by RANS, while only the actual separated flow portion should be *simulated* by LES [15]. The division between the modelled and simulated zones is based however on the RANS capabilities in foreseeing the separation point, a very crucial aspect especially at low Reynolds number where bubble formation takes place on airfoils. An improvement of this weakness would imply the presence of the same

LES informations in the upstream flow that would lead to an undesired increase in time-duration for the computations.

As depicted by Spalart in the first work of 1997 [14] (referred as DES97) the switching between the RANS and LES is based on a DES length scale defined as

$$L_{DES97} = \min(d_w, L_{LES}),\ L_{LES} = C_{DES}\Delta,\ \Delta = \max(\Delta_i, \Delta_j, \Delta_k) \tag{1}$$

where d_w is the wall-normal distance, L_{LES} the LES length scale, C_{DES} a DES model constant and Δ corresponds to a LES width filter related to the maximum cell size in respect to all the directions. The constant C_{DES} can be regarded similarly to the Smagorinsky constant [9] and its first calibrated value was set equal to 0.65.

The simulations carried out in this work refer to an enhanced version of DES, the so called DDES (Delayed Detached-Eddy Simulation) proposed again by Spalart in 2006 [16].

The DDES version suggested in [16] is the one currently implemented in OpenFOAM, which has been used for this simulations.

Particular care is given to the turbulence quantities settings for the inlet boundary as pointed out by Spalart [17]. In order to achieve a turbulence intensity comparable with that of the experiments, the free decay of the turbulence quantities from the inlet to the airfoil should be taken into account. Considering the solution of $k - \varepsilon$ equations and assuming that when approaching the body only the destruction terms are active, one gets the solution for the turbulent viscosity

$$\nu_t = \nu_{tI} \left[1 + (C_{\varepsilon 2} - 1) \left(\frac{\varepsilon}{k} \right)_I \frac{x}{U} \right]^{\frac{C_{\varepsilon 2} - 2}{C_{\varepsilon 2} - 1}}, \tag{2}$$

from which one can find the modified turbulent viscosity at the inlet using the S–A model relation [13].

The presence of a laminar flow region on the airfoil has also been considered, since the Reynolds number falls in the so called *low-Re* regime (Re $\lesssim 10^6$). This has been done using an *low-Re* incompressible RANS model, implemented in the OpenFOAM toolbox 2.1.0, for flows with a boundary layer that undergoes a transition from laminar to turbulent. The model is described in the work of Walters [21].

3 Simulation Mesh

The adopted grid consists of a 3-D full block-structured mesh made with the OpenFOAM [10] native mesher *blockMesh*. The trailing edge of the airfoil has been rounded in order to guarantee a high quality O-Type mesh [9, 4, 7], avoiding at the same time the need of using an intermediate C-Grid. The distance between the inlet/outlet boundary and the airfoil is 15 unit chords , while the mesh spanwise length is 1 unit chord (no solution is provided for the third component of the velocity field). The chord length is 0.2 m, exactly as it was in the experiment of Schneemann [11]. The first cell layer around the airfoil has a wall normal distance such that $y^+ \approx 1$

for a Reynolds Re = 700000. A wall function is therefore not required. The overall number of cells is 173218, while on the airfoil there are 337 nodes.

The inlet boundary condition for the modified turbulent viscosity $\tilde{\nu}_t$ is based on the suggestions from Spalart [17].

4 Results

The simulations carried out within this work are compared with measurements performed in the closed test section of the closed loop wind tunnel of Oldenburg University ($1.0 \times 0.8\,\mathrm{m}^2$, 2.0 m long) [11]. Reynolds numbers Re $\simeq$ 700000 were reached (chord length of the vertically mounted aluminum milled airfoil $c = 0.2\,\mathrm{m}$, wind speed $u \simeq 50\,\mathrm{m/s}$). In the experiment, the turbulent inflow condition is generated via a rectangular grid installed on the nozzle of the wind tunnel with grid width of 50 mm.

A comparison of the modified turbulent viscosity decay between the theoretical model and the numerical result has been performed. The modified turbulent viscosity in the flowfield has been normalized with its value at the body $\tilde{\nu}_{t,\mathrm{exp}}$ derived from the turbulence intensity that was experimentally measured. This procedure is adopted for the theoretical model and for the numerical result. Figure 1 shows the turbulent viscosity decaying with the presence of the body.

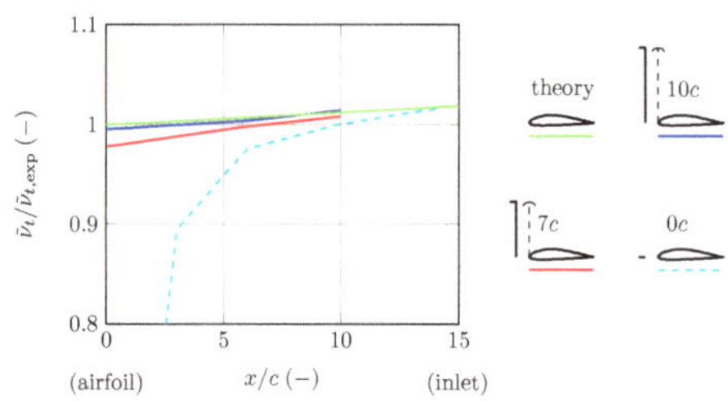

Fig. 1 Theoretical vs. numerical decay for $Ti = 3.6\%$: — theory Eq. 2, — simulated probe 10 chords ahead from the airfoil, — simulated probe 7 chords ahead from the airfoil, - - simulated probe at the airfoil

Figure 2 shows a comparison between measured and computed lift coefficient for the turbulence inflow condition such that $Ti = 3.6\%$. The simulations include a Spalart-Allmaras RANS (indicated as SARANS), a Spalart-Allmaras DDES (indicated as SADDES), a $k - k_L - \omega$ RANS and a $k - k_L - \omega$ URANS simulations.

It can be easily observed that only the simulations performed with the $k - k_L - \omega$ transitional model are able to predict satisfactorily the linear range of the lift curve. This behavior is attributed to the lack of a laminar flow region in the other turbulence models. Therefore, the authors would like to highlight the importance of considering a turbulence transition model, which at least for the linear part of the curve has proved to be more crucial than the use of a RANS-LES hybrid approach.

Concerning the stall regime, the SARANS model only obtains the trend behavior (the same happens with SAURANS, not shown here). A better description is obtained from the $k - k_L - \omega$ URANS model (the steady solution is not as good as the

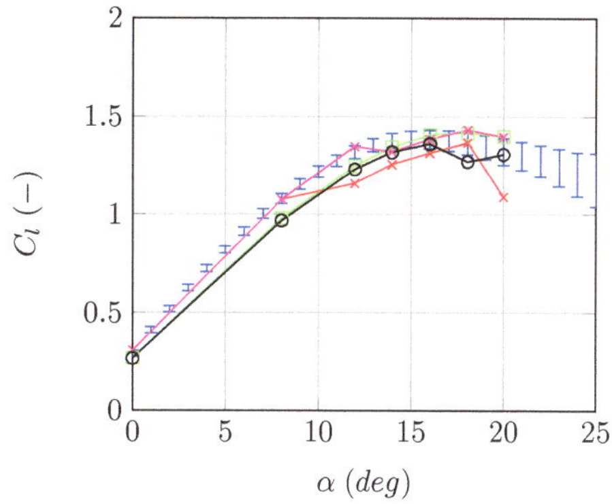

Fig. 2 I Experiment, SARANS, SADDES, $k-k_L-\omega$RANS , $k-k_L-\omega$URANS

unsteady), providing satisfactory results which fit with the error bars of the experiment. The SADDES solution shows different stall characteristics than SARANS due to LES effects, even if the linear range also suffers from the absence of a laminar contribution.

5 Conclusions and Future Works

The turbulence decay within the simulation domain has been studied in order to prescribe the proper turbulence settings in the inlet. The scope of having the desired turbulence level reaching the studied body has been be satisfactorily achieved thanks to the decay equations provided by Spalart [17]. The presence of the body itself is then responsible for the re-laminarization of the flow.

The use of a turbulence transition model [21] results in a good prediction capability of the URANS simulations (even for the 2D case). Therefore, the more time-consuming 3D simulations (as proposed in [1]) can be avoided.

The lack of the transition model for DDES makes this kind of simulation not so reliable as URANS (at least for the linear part of the lift polar). A future implementation of the mentioned transition model in the DDES solver of OpenFOAM is expected to improve substantially its capability for predicting *low-Re* flows.

Furthermore the authors will analyze in the next future the behavior of the FX79W151A airfoil in a turbulent inflow induced by a fractal grid. The simulation results will be compared with already existing measurements [11].

Further simulations with still lower Reynold numbers are also planned for their validation with available PIV data.

Finally, the authors are aiming at verifying with other measurements [22] the reliability of the DDES approach for studying dynamic stall phenomena.

References

1. Bertagnolio, F., Sørensen, N., Johansen, J.: Profile catalogue for airfoil sections based on 3D computations. In: Risø-R-1581(EN) (2006)
2. Davidson, L., Cokljat, D., Fröhlich, J., Leschziner, M.A., Mellen, C., Rodi, W.: LESFOIL: Large Eddy Simulation of Flow Around a High Lift Airfoil. In: Notes on Numerical Fluid Mechanics, vol. 83. Springer (2003)
3. Gilling, L., Sørensen, N.N., Davidson, L.: Detached Eddy Simulations of an Airfoil in Turbulent Inflow. In: 47th AIAA Aerospace Sciences Meeting Including The New Horizons Forum and Aerospace Exposition, Orlando, Florida, January 5-8 (2009)
4. Haase, W., Braza, M., Revell, A.: DESider - A European effort on hybrid RANS-LES modelling. In: Notes on Numerical Fluid Mechanics and Multidisciplinary Design, vol. 103. Springer, Berlin (2009)
5. Johansen, J., Sørensen, N.: Application of a Detached-Eddy Simulation model on airfoil flows. In: 14th Symposium IEA Joint Action, Aerodynamics of Wind Turbines, Boulder, CO (December 2000)
6. Li, D.: Numerical simulation of thin airfoil stall by using a modified DES approach. International Journal for Numerical Methods in Fluids 54(3), 325–332 (2007), doi:10.1002/fld.1403.
7. Madsen, J., Lenz, K., Dynampally, P., Sudhakar, P.: Investigation of grid resolution requirements for detached eddy simulation of flow around thick airfoil sections. In: EWEC 2009, Parc Chanot, Marseille, France, March 16-19 (2009)
8. McGhee, J.R., Walker, B.S., Millard, B.F.: Experimental results for the Eppler 387 Airfoil at low Reynolds numbers in the Langley low-turbulence pressure tunnel. In: NASA Technical Memorandum 4062 (1988)
9. Mockett, C.: A comprhensive study of detached-eddy simulation. Ph.D. thesis, Institute of Fluid Mechanics and Engineering Acoustics, Technische Universität Berlin (2009)
10. `http://www.openfoam.org`
11. Schneemann, J., Knebel, P., Milan, P., Peinke, J.: Lift measurements in unsteady flow conditions. In: EWEC 2010, Warsaw, Poland, April 20-23 (2010)
12. Somers, D.M., Tangler, J.: Design and experimental results for the S809 Airfoil. NREL/SR-440-6918 UC Category:1213 D 97000206, 1–104 (1997)
13. Spalart, P.R., Allmaras, S.R.: A one-equation turbulence model for aerodynamic flows. Recherche Aerospatiale (1), 5–21 (1994)
14. Spalart, P.R., Jou, W., Strelets, M., Allmaras, S.: Comments on the feasibility of LES for wings, and on a hybrid RANS/LES approach. In: Advances in DNS/LES, vol. 1 (1997)
15. Spalart, P.R.: The uses of DES: natural, extended, and improper. Invited presentation at the DESider Hybrid RAN-LES Symposium, Stockholm, Sweden, July 14-15 (2005)
16. Spalart, P.R., Deck, S., Shur, M., Squires, K., Strelets, M., Travin, A.: A new version of detached-eddy simulation, resistant to ambiguos grid densities. In: Theoretical and Computational Fluid Dynamics, vol. 20, pp. 181–195 (2006)
17. Spalart, P.R., Rumsey, C.L.: Effective Inflow Conditions for Turbulence Models in Aerodynamic Calculations. AIAA Journal 45(10), 2544–2553 (2007)
18. Spalart, P.R.: Detached-eddy simulation. Annual Review of Fluid Mechanics 41, 181–202 (2009)
19. Tangler, J., Somers, D.M.: NREL Airfoil Families for HAWTs. In: AWEA, pp. 1–12 (1995)

20. Timmer, W.A.: Two-dimensional low-Reynolds number wind tunnel results for airfoil NACA 0018. Wind Engineering Volume 32(6), 525–537 (2008)
21. Walters, D.K., Cokljat, D.: A Three-Equation Eddy-Viscosity Model for Reynolds-Averaged Navier Stokes Simulations of Transitional Flow. J. Fluids Eng. 130(12), 121401, 14 (2008), doi:dx.doi.org/10.1115/1.2979230
22. Wolken–Möhlmann, G., Knebel, P., Barth, S., Peinke, J.: Dynamic lift measurements on a FX79W151A airfoil via pressure distribution on the wind tunnel walls. The Science of Making Torque from Wind. Journal of Physics: Conference Series 75, 012026 (2007), doi:10.1088/1742-6596

Single and Multi-element Airfoil Performance Simulation Study and Wind Tunnel Validation

O. Eisele and G. Pechlivanoglou

Abstract. The focus of the present investigation is to evaluate the capability of modern state of the art codes to simulate the aerodynamic behavior of typical wind turbine airfoils. Two airfoils were selected. The AH 93-W-174 [2] with a thickness of 17.4% as a representative for relatively thin airfoils in the outer region and the relatively thick DU 97-W-300 [9] with a thickness of 30% as a representative for rather thick airfoils in the inner region of wind turbine blades. The third testcase was a multi-element configuration composed of the DU-97-W-300 as the main airfoil and a slat based on the NACA 22 airfoil. For all configurations aerodynamic simulations with different levels of complexity were accomplished. First simulations were carried out with the panel code XFOIL [3] and the Euler code MSES [4]. Furthermore steady state RANS computations were conducted by the use of the open source code OpenFOAM® [6]. There the Spalart Allmaras [8], the two-equation $k-\omega$ shear stress transport (SST) model by Menter [5] , and the recently implemented three-equation $k-k_l-\omega$ turbulence models by Walter and Cokljat [10] were applied. All simulation results are compared with wind tunnel data, measured at the large wind tunnel facilities of HFI/ISTA TU-Berlin.

1 Introduction

In order to accurately model the aerodynamic behavior of wind turbine blades by the means of power extraction from the wind and resulting blade loads, the aerodynamic characteristics of the airfoils that the blade is composed of are required. Lift and drag data for airfoils can be either measured in a wind tunnel or computed numerically. Nowadays, carefully conducted wind tunnel measurements generally give the most reliable results but are very cost intensive. Since the computational

O. Eisele · G. Pechlivanoglou
Smart Blade GmbH, Warschauer Str. 38, 10243 Berlin
e-mail: {o.eisele,g.pechli}@smart-blade.com

M. Hölling et al. (eds.), *Wind Energy – Impact of Turbulence*,
Research Topics in Wind Energy 2,
DOI: 10.1007/978-3-642-54696-9_3,

power has grown in the recent years, computational fluid dynamic methods are also incorporated into modern blade design processes. However, these methods contain assumptions and one has to be critical when using the resulting data. The aim of the research is to compare the results of state of the art fluid dynamic codes with wind tunnel measurements of several configurations. All the wind tunnel measurements were performed at the large wind tunnel of H.F.I TU Berlin at a Reynolds number of $Re = 1.3 \times 10^6$ a turbulence level of about $Tu_\infty = 0.1\%$. More information about the wind tunnel facilities and the testing procedure can be found in the disertation of Pechlivanoglou [1].

2 Test Case Selection

Wind turbine blades are composed of thick airfoils in the inner region for structural integrity and thin airfoils in the outer region for aerodynamic performance. Furthermore, research has shown that auxiliary devices like slats on wind turbine blades can lead to an increase of the blade performance [1]. In order to account for that, two different airfoils as well as an additional multi-element configuration , shown in Fig. 1, were selected for the investigation. The first configuration is the AH 93-W-174 with a relative thickness of 17.4%, a slight camber and a relatively flat pressure side with an s-shaped trailing edge. The DU 97-W-300 airfoil has a relative thickness of 30% and a strong s-shaped pressure side. As a third configuration the DU 97-W-300 was combined with a slat based on the NACA 22 airfoil. The chord of the slat is 25% of the airfoil chord and the position was chosen according to reviously conducted wind tunnel test [1].

3 Simulation Tools and Setup

For a numerical computation of the airfoil performance three different codes with different levels of complexity were applied. The panel code XFOIL, the Euler code

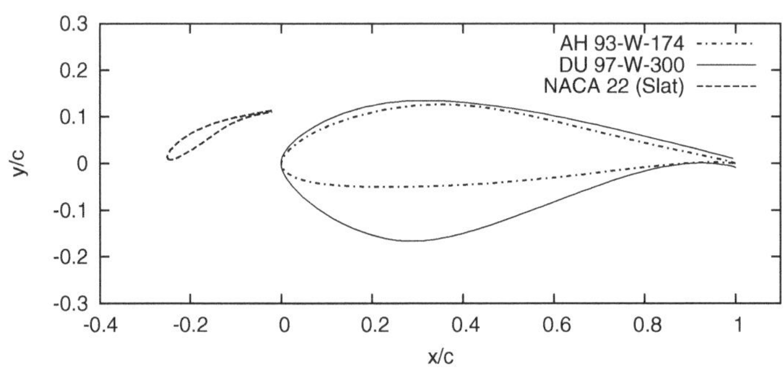

Fig. 1 Coordinates of the test configurations, the AH 93-W-174, the DU 97-W-300 and the multi-element configuration DU 97-W-300 combined with NACA 22

MSES and the open source CFD-toolbox OpenFOAM®. All simulations were carried out with a Reynolds number of $Re = 1.3 \times 10^6$, which was the same as for preliminary conducted wind tunnel measurements. In the following a brief description of the codes as well as the simulation setups are presented.

3.1 XFOIL

The aerodynamics of airfoils are computed by XFOIL [3] based on a panel method combined with an integral boundary layer formulation. Transition is predicted by the semi-empirical e^N envelope method. The value of $N = 9$ (critical amplification) was selected based on several measurements of A. M. O. Smith [7], which showed that this value leads to a satisfactory determination of the transition location .

The aerodynamics of the AH 93-W-174 airfoil and the DU 97-W-300 airfoil were computed with the use of 200 panels distributed along the airfoil surface. The critical amplification was set to e^9 for the AH 97-W-174 and to e^2 for the DU 97-W-300 airfoil. XFOIL has no capability for the computation of multi-element configurations.

3.2 MSES

The code MSES [4] of Drela solves the Euler equations on a discrete 2D grid coupled with an integral boundary layer formulation. Transition is predicted by the same method as described in Sec. 3.1. Grid generation and post processing functionality is included in the code.

For all configurations grids with approx. 6000 cells were created, whereas the distance of the farfield boundaries was set at 10 chord lengths around the airfoil. For the AH 93-W174 a transition criterion of e^9 was used and for the DU 97-W-300 and the multi-element configuration a transition criterion of e^2 was used for the same reasons as mentioned in Sec. 3.1.

3.3 OpenFOAM®

Steady state Reynolds-Averaged Navier-Stokes (RANS) computations were accomplished by the use of the simpleFOAM solver of the open source CFD-toolbox OpenFOAM® [6]. The single element configurations were meshed with a block structured O-Grid. The multi-element configuration was meshed with a hybrid grid, which is composed of structured boundary layer grids for each element and an unstructured grid for the rest of the domain. In order to resolve the viscous sublayer of the boundary layer the dimensionless wall distance was $y^+ < 1$ along the entire airfoil contour. A second grid with a dimensionless wall distance of $y^+ > 15$ was created by coarsening the grid normal to the airfoil in order to be able to run simulations were the boundary layer is not resolved, but is described instead by the use of wall functions. The number of cells for all grids was in the range of 10^5.

The turbulence models which were applied for the simulations were, the one-equation Spalart Allmaras [8] turbulence model combined with a wall function for the boundary layer, the two-equation Shear Stress Transport (SST) model by Menter [5] and the recently implemented $k-k_l-\omega$ model by Walters and Cokljat [10] both with resolved boundary layer. At the outer boundaries of the domain the quantities of the turbulence models, i.e the turbulent kinetic energy k and the inverse turbulent time-scale ω, were computed based on a turbulent viscosity ratio of $\mu_T/\mu = 1$ and a turbulent intensity of $Tu_\infty = 0.1\%$.

4 Results

The simulation results for the AH 93-W-174 airfoil and the DU 97-W-300 airfoil are presented in Fig. 2 and Fig. 3. Regarding AH 93-W-174 both codes, XFOIL and MSES, predict similar lift and drag characteristics. Compared to the experiment the lift and the stall characteristics of the airfoil are accurately captured. However, drag is significantly underestimated. In case of the relatively thick DU 97-W-300 airfoil MSES has the same behavior as for the AH 93-W-174 in terms of accurate lift prediction but underestimation of drag. XFOIL however, fails to model the aerodynamic behavior of this airfoil and leads to a significant over-prediction of the maximum lift coefficient $c_{l,max}$.

The results of the RANS computations for the AH 93-W-174 and the DU 97-W-300 show a similar trend for all turbulence models. Pre- and post stall behavior in terms of drag and lift are in good agreement with the wind tunnel measurements. However, all accomplished RANS computations tend to under-predict the maximum lift $c_{l,max}$ in the angle of attack region between 7° and 12° where the aerodynamic

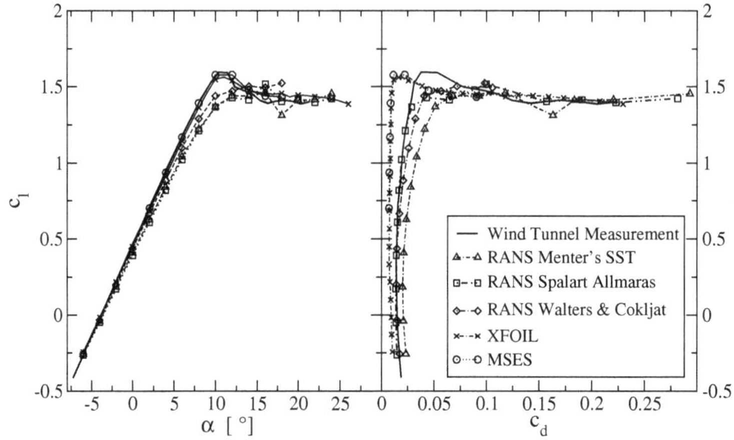

Fig. 2 Comparison of simulation results for the AH 93-W-174 airfoil. Left) lift coefficient vs. angle of attack, Right) lift coefficient vs. drag coefficient.

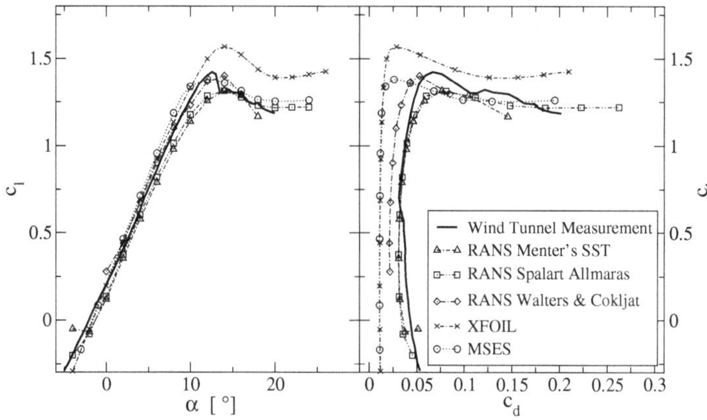

Fig. 3 Comparison of simulation results for the DU 97-W-300 airfoil. Left) lift coefficient vs. angle of attack, Right) lift coefficient vs. drag coefficient.

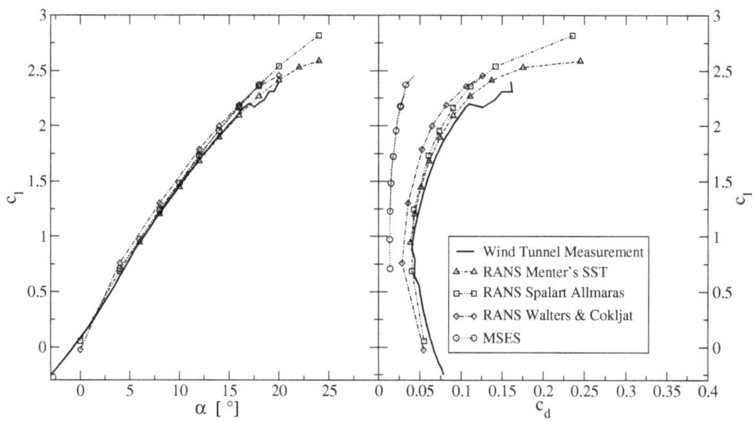

Fig. 4 Comparison of simulation results for the multi-element configuration. Left) lift coefficient vs. angle of attack, Right) lift coefficient vs. drag coefficient.

behavior of the airfoils is significantly dominated by the onset of stall. In case of the DU 97-W-300 airfoil, however, in terms of the prediction of $c_{l,max}$ the turbulence model of Walters and Cokljat performs best.

The multi-element simulations (Fig. 4) accurately predict the lift characteristics of the configuration up to an angle of attack of 17°. Beyond this angle the slat starts to stall, which is not rendered by the simulations. As in the previous cases MSES significantly under-predicts drag, whereas the steady RANS simulations result in relatively accurate drag predictions.

5 Conclusion

Both XFOIL and MSES, underestimate the drag in the pre-stall regime, which is mainly driven by the shear friction of the boundary layer. Therefore, this seems to be a drawback of their integral boundary layer formulation. The use of RANS simulations with a detailed boundary layer modeling is a good alternative. Besides slight differences, all tested CFD models show a similar trend in the results and a good agreement with the wind tunnel measurements. Hence, the careful use of CFD methods seems to be a promising alternative to cost intensive wind tunnel measurements.

References

1. Pechlivanoglou, G.: Passive and active flow control solutions for wind turbine blades. OPUS TU, Berlin (2013)
2. Althaus, D.: Niedriggeschwindigkeitsprofile: Profilentwicklungen und Polarenmessungen im Laminarwindkanal des Instituts für Aerodynamik und Gasdynamik der Universität Stuttgart. Vieweg, Braunschweig (1996)
3. Drela, M.: XFOIL: An Analysis and Design System for Low Reynolds Number Airfoils. In: Lecture Notes in Engineering: Low Reynolds Number Aerodynamics. Springer, Notre Dame (1989)
4. Drela, M.: A User's Guide to MSES 3.04. MIT Department of Aeronautics and Astronautics
5. Menter, F.R.: Improved Two-Equation $K - \omega$ Turbulence Models for Aerodynamic Flows. In: NASA STI/Recon Technical Report N (1992)
6. OpenFOAM, The Open Source CFD Toolbox, `http://www.openfoam.com` (May 14, 2012)
7. Smith, A.M.O.: Transition pressure gradient and stability theory. Journal of the Aero/Space Sciences 26 (1959)
8. Spalart, P.R., Allmaras, S.R.: A One-Equation Turbulence Model for Aerodynamic Flows. La Recherche Aerospatiale (1) (1997)
9. Timmer, W.A., Van Rooij, R.P.J.O.M.: Summary of the Delft University wind turbine dedicated airfoils. In: Journal of Solar Energy Engineering (2003)
10. Walters, D.K., Cokljak, D.: A Three-Equation Eddy-Viscosity Model for Reynolds-Averaged Navier-Stokes Simulations of Transitional Flows. ASME Journal of Fluids

Pairing Instability in Helical Vortices

Hadrien Bolnot, Stéphane Le Dizès, and Thomas Leweke

1 Introduction

The stability properties of helical vortices are of interest for applications such as wind turbine and helicopter wakes. In the former, they contribute to the transition from an organised vortex system to a turbulent wake behind the rotor, and in the latter they may be related to the inception of the so-called Vortex Ring State of the wake of a helicopter in steep descent [2, 7]. In this work, we consider the particular instability that leads to vortex pairing, for which previous experimental [1, 4] and numerical [5] studies have shown that it plays an important role in rotor wakes. In the following, we recall a few theoretical results concerning this phenomenon, and then present experimental observations of pairing in a single helical vortex under carefully controlled conditions.

2 Pairing Instability of Vortex Arrays

Vortex pairing is a process occurring in (infinite) arrays of identical concentrated vortices, whereby small perturbations of their initially equidistant positions are amplified in a way that neighbouring vortices approach each other and group in pairs. It occurs, e.g., in shear layers, as a secondary instability of the Kelvin-Helmholtz instability. Pairing is distinct from merging of two vortex cores of like-signed vorticity into a single one, which occurs when the two initial vortices come sufficiently close to each other (see, e.g., [9]). Merging may take place during the late stages of the pairing instability in arrays of real (distributed) vortices.

Systems of helical vortices, such as those found in the wake of a rotor, present locally arrays of identical curved vortices, which are expected to exhibit the pairing

Hadrien Bolnot · Stéphane Le Dizès · Thomas Leweke
IRPHE UMR 7342, CNRS, Aix-Marseille Université, 13384 Marseille, France
e-mail: Thomas.Leweke@irphe.univ-mrs.fr

M. Hölling et al. (eds.), *Wind Energy – Impact of Turbulence*,
Research Topics in Wind Energy 2,
DOI: 10.1007/978-3-642-54696-9_4, © Springer-Verlag Berlin Heidelberg 2014

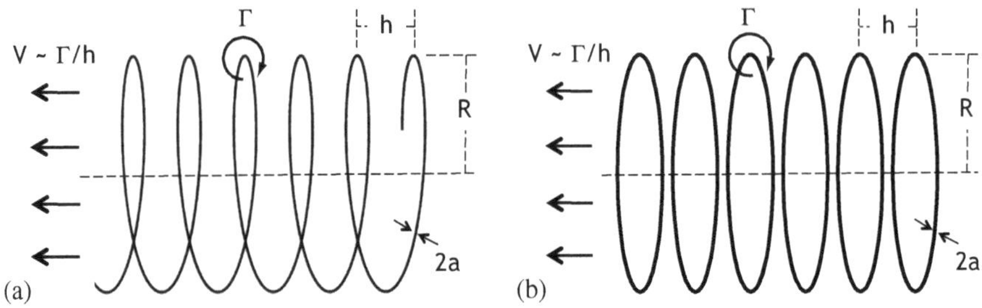

Fig. 1 Geometry and parameters of (a) a helical vortex filament and (b) an array of vortex rings

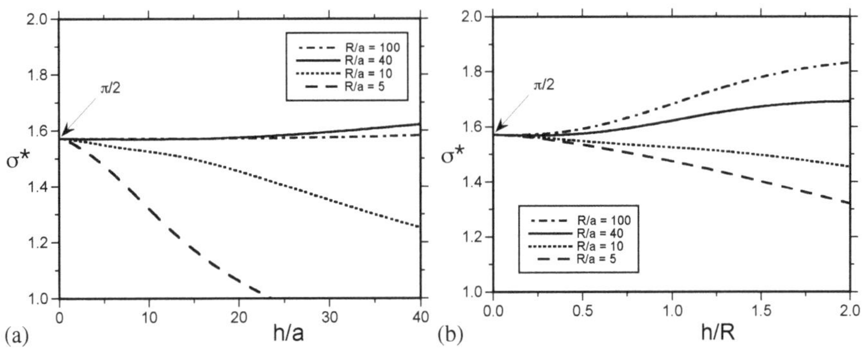

Fig. 2 Growth rate of the pairing instability for an array of vortex rings, as function of the separation h normalised by (a) the core size and (b) the helix radius (after [8])

instability. The simplest system consists in a single helical vortex filament, characterised by its circulation Γ and core radius a, as well as the helix radius R and pitch h (Fig. 1a). In the frame of reference where the fluid at infinity has no velocity in the direction of the helix axis, the fluid inside the helix moves at a speed of the order of Γ/h, and the helix itself moves with a velocity of roughly half this value ($\Gamma/2h$). The latter velocity and h are used here for non-dimensionalisation.

For small pitch ($h \ll R$), the helix geometry is locally very similar to that of an array of axisymmetric vortex rings (Fig. 1b). Levy & Forsdyke [8] have treated analytically the stability of such a system with respect to pairing of neighbouring vortex rings. Figure 2 shows the non-dimensional growth rate $\sigma^* = \sigma \cdot (2h^2/\Gamma)$ of the pairing instability as function of the geometric parameters. These were calculated using the same procedure as in [8], including also the effects of a smooth (Gaussian) vorticity distribution and of an axial core flow (see Sect. 3, Fig. 4b), as well as the variation of core size through stretching. For small separation/pitch ($h \ll R$) and small core size ($a \ll R$), which includes most configurations relevant for applications, the growth rate approaches the value $\pi/2$, i.e., the growth rate of the pairing instability for a single [3] or double [6] row of point vortices in two dimensions.

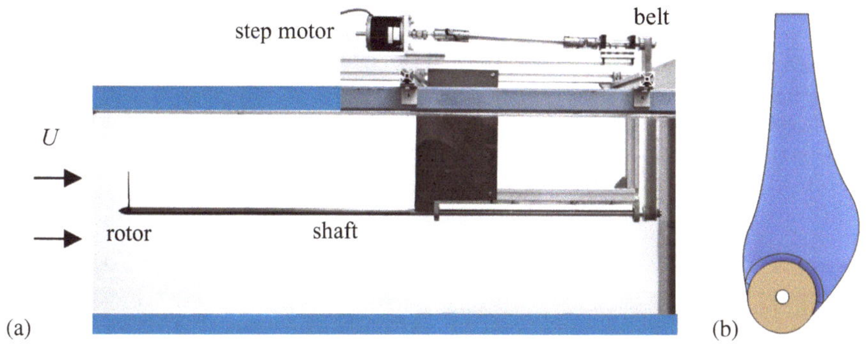

Fig. 3 (a) Side view of the water channel test section, showing the set-up used to generate a single helical vortex. (b) Plan view of the blade geometry.

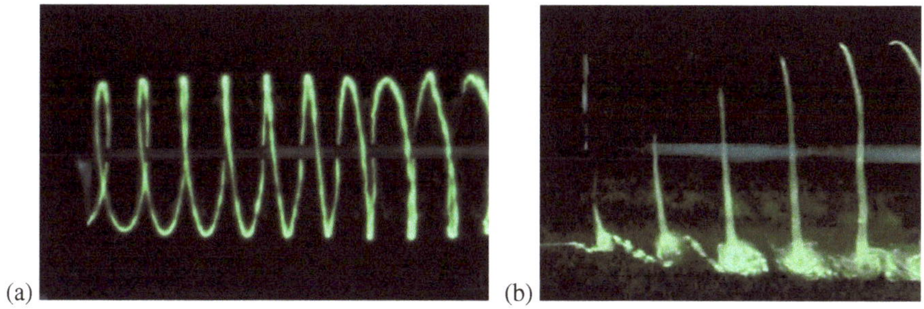

Fig. 4 Dye visualisation of the unperturbed helical vortex with $h/R = 0.53$ at $Re_\Gamma = 11500$. (a) Dye washed off the blade tip; (b) dye injected at a fixed position near the edge of the rotor disk.

3 Experiments

Experiments on the controlled pairing of a single helical vortex were carried out in a recirculating free-surface water channel with a test section of dimensions 38 cm (width) × 50 cm (height) × 150 cm (length). The vortex was generated near the test section entry by a single-bladed rotor mounted on a shaft and driven by a computer-controlled stepper motor outside the test section using a belt (Fig. 3a). The rotor blade geometry (Fig. 3b) is based on the low-Reynolds number airfoil A18 by Selig *et al.* [11], with chord and twist distributions designed to operate in the wind turbine regime and produce a constant radial circulation distribution (Joukowsky rotor, see e.g. [10]) over the outer 75% of the span, in order to generate a highly concentrated tip vortex. The rotor has a radius $R_o = 80$ mm and a tip chord $c = 10$ mm. For the present set of experiments, it is rotated at a frequency $f = 6$ Hz and placed in a uniform flow with a free stream velocity $U = 36$ cm/s, resulting in a tip speed ratio $\lambda = 8.4$. The tip chord Reynolds number is $Re_c = 2\pi R_o f c/\nu = 30000$, and the one based on the vortex circulation Γ (determined from Particle Image Velocity measurements) is $Re_\Gamma = \Gamma/\nu = 11500$. ν is the kinematic viscosity.

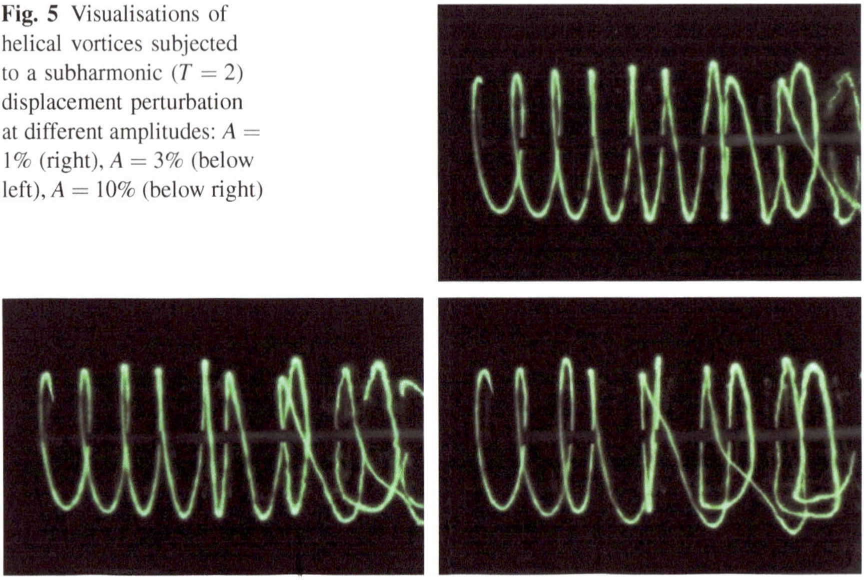

Fig. 5 Visualisations of helical vortices subjected to a subharmonic ($T = 2$) displacement perturbation at different amplitudes: $A =$ 1% (right), $A = 3\%$ (below left), $A = 10\%$ (below right)

The helical vortex structure is visualised using fluorescent dye (fluorescein) washed off the blade tip and illuminated by the light of an argon ion laser. Figure 4 shows the vortex produced for the present set of conditions, it is regular and unperturbed for about 10 helix turns. Dye injection at a fixed location near the blade tip trajectory (Fig. 4b) reveals the presence of a strong flow inside the core along the vortex axis. The dye pattern visualises the corresponding axial velocity profile and allows an estimate of the peak velocity V_a and the vortex core diameter $2a$ (width of the profile at $1/e$ of its maximum). The helix parameters are here found as: $h/R = 0.53$, $h/a = 20$, $R/a = 38$, $V_a = 0.4(\Gamma/2\pi a)$.

Pairing of successive helix loops was then induced by a controlled perturbation of the blade rotation, resulting in a varying streamwise displacement of the vortex. The modified helix geometry can be expressed as:

$$z/h = \theta/2\pi + A\cos(\theta/T), \quad (1)$$

where z and θ are the downstream and azimuthal positions, and A the relative displacement amplitude. For $T = 2$, successive helix loops are displaced in opposite directions, with an amplitude varying in the azimuthal direction, triggering local pairing. Figure 5 shows the vortex structure for three different amplitudes of forcing, with the maximum displacement occurring at the top (the perturbation vanishes at the bottom). A displacement of as little as 1% of h is seen to trigger the pairing instability, leading to a grouping and 'swapping' of successive loops as the helix moves downstream. The azimuthal variation of the pairing results in a complicated three-dimensional vortex structure, but no merging or breakdown to small-scale structures is seen. As the amplitude is increased, the deformations appear closer to the rotor, as expected.

Fig. 6 Measurements of the amplitude d (in arbitrary units) of the pairing perturbation as function of downstream distance from visualisation sequences of the cases shown in Fig. 5

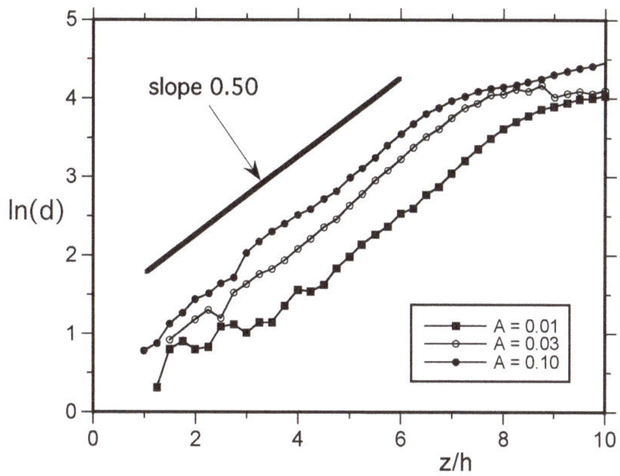

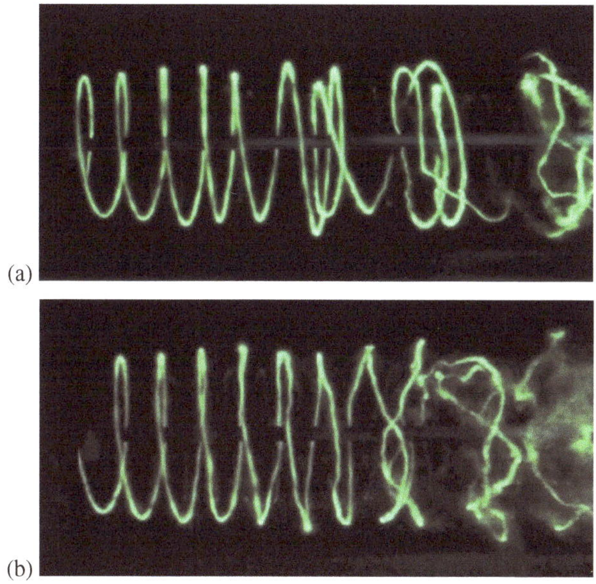

Fig. 7 Examples of more general displacement perturbations leading to instability. (b) Tripling ($T = 3$ and $A = 3\%$); (b) Short-wave perturbation ($T = 1/4$, $A = 10\%$).

The growth rate of the pairing instability can be estimated from the evolution of the displacement d of the vortex, with respect to the unperturbed position, which can be measured from video sequences of visualisations such as in Fig. 5. The result for these three cases is plotted in Fig. 6; the spatial growth rate σ_s is found to be close to $0.5/h$. Using the (measured) convection velocity V_c of the helix, one can obtain an estimate of the temporal growth rate $\sigma = \sigma_s V_c$, leading finally to a non-dimensional growth rate $\sigma^* = 1.2$ for these experiments. This is about 25% lower than the theoretical prediction ($\sigma^* \approx \pi/2$) for a row of vortex rings with the same geometrical parameters (Fig. 2). The difference may be due to the non-uniform nature of the pairing for a single helix. The effect of the central hub vortex, always present in the wake of a rotor, also needs to be clarified. Further experiments involving two helical vortices (two-bladed rotor), where pairing can occur globally, are underway.

Vortex perturbations of the form given in Eq. (1) can be used to trigger various other types of 'pairing'. Two examples are shown in Fig. 7: a perturbation of period $T = 3$, leading to a grouping of three successive helix loops, and a perturbation on a shorter length scale ($T = 1/4$, four wavelengths in one turn). The latter case is seen to exhibit a breakdown of the vortex structure, triggered by local pairing and merging. It bears striking similarities with the numerical simulation results obtained by Ivanell *et al.* [5].

4 Conclusion

The experimental observation of the dynamics of a single helical vortex under carefully controlled conditions has shown that this flow is highly receptive to a variety of displacement perturbations leading to pairing of successive helix loops or more complicated three-dimensional deformations involving local merging and breakdown. These findings are coherent with the results from previous experimental studies of rotor wakes (e.g., [1, 4]), where helix loop grouping is consistently observed in a way which reflects the symmetry of the set-up (pairing for two blades, tripling for three blades, etc.), suggesting that this phenomenon might be triggered by small asymmetries in the rotor geometry. The great sensitivity of the rotor wake to displacement perturbations, and the fact that their amplitude influences the distance over which the helical wake evolves, may add further insight into the effect of rotor vibrations or external turbulence on the development of wind turbine wakes, and potentially lead to ideas for their control.

Acknowledgement. This work is supported by EUROCOPTER S.A.S., under contract no. IPEC 0187N/2009.

References

1. Alfredsson, P.H., Dahlberg, J.-Å.: Technical Note AU-1499 (Part 7), The Aeronautical Research Institute of Sweden, Stockholm (1979)
2. Bolnot, H., Le Dizès, S., Leweke, T.: AIAA Paper 2011-3927 (2011)
3. Branger, P., Chomaz, J.M.: Phys. Rev. Lett. 78, 658 (1997)
4. Felli, M., Camussi, R., Di Felice, F.: J. Fluid Mech. 682, 5 (2011)
5. Ivanell, S., Mikkelsen, R., Sørensen, J.N., Henningson, D.: Wind Energy 13, 705 (2010)
6. Lamb, H.: Hydrodynamics, 6th edn. Cambridge University Press (1932)
7. Leishman, J.G., Bhagwat, M.J., Ananthan, S.: J. Am. Helicopter Soc. 49, 160 (2004)
8. Levy, H., Forsdyke, A.G.: Proc. R. Soc. Lond. A 114, 594 (1927)
9. Meunier, P., Le Dizès, S., Leweke, T.: C. R. Physique 6, 431 (2005)
10. Okulov, V.L., Sørensen, J.N.: J. Fluid Mech. 649, 497 (2010)
11. Selig, M.S., Guglielmo, J.J., Broeren, A.P., Giguere, P.: Summary of Low-Speed Airfoil Data, SoarTech, Virginia Beach, vol. 1 (1995)

Actuator Disc Modeling of the MEXICO Rotor Experiment

Esmail Mahmoodi and Alois Peter Schaffarczyk

Abstract. A differential Actuator Disc (AD) based on 2D airfoils data has been prepared for the so called MEXICO rotor experiment. A commercial Navier-Stokes-Solver (FLUENT 5.7) without any turbulence modeling has been used. The geometrical model includes both the full DNW Large Low speed Facility (LLF) wind- tunnel geometry where the experiment was performed as well as a cylindrical domain used for the boundary-free case, to estimate the blockage effects of the wind tunnel. Results have been prepared for rotor performance, loads and velocity fields in the near-wake which were measured for the first time during the experiment and are also compared to an RANS model.

1 Introduction

Modeling Wind-Turbine Flow may be studied in levels of varying sophistication: The Blade Element-Momentum (BEM) theory has been used since almost hundred years now being replaced by CFD models of RANS kind, which are exposed to the usual difficulties concerning the turbulent and transitional part of the flow necessary to reproduce the high lift-to-drag ratios on the blade. As an intermediate step simplified Actuator Disc (AD), Actuator Line (AL) and Actuator Surface (AS) models are used instead. To mention only few of them who applied these methods to the MEXICO Experiment (described in [1]) and the IEAwind Mexnext Project: BEM model by [2], the RANS model by [3], the AL model by [4], the AS by [5].

Esmail Mahmoodi
College of Engineering and Technology, University of Tehran, Karaj, Iran
e-mail: esmail.mahmoodi@gmail.com

Alois Peter Schaffarczyk
Mechanical Engineering Department, University of Applied Sciences Kiel, Kiel, Germany
e-mail: Alois.schaffarczyk@fh-kiel.de

M. Hölling et al. (eds.), *Wind Energy – Impact of Turbulence*,
Research Topics in Wind Energy 2,
DOI: 10.1007/978-3-642-54696-9_5,

Unlike others [6] used loads from measurements to explain the so far un-explained differences in simulated flow field from the measured ones.

2 Methods

The actuator disc was meshed using ICEM CFD software using O-type topology. Only structured meshes were used. The number of cells at the disc was: 36 in radial, 72 in circumferential and finally 2 in axial direction (Fig.1a). A hybrid mesh has been used for meshing the wind tunnel (Fig.1b). A Quad-dominant mesh was used to prepare the surface grids and the Tetra-mixed method was used for the volume part of the geometry. The disc then was merged inside the wind tunnel mesh to prepare the model for full wind tunnel and results in 6.08 Mcells and 1. Mnode resp. For boundary-less flow, a cylindrical O-type (Fig.1c) was modeled. In the mesh had about by 1.04M cells and 1.06M nodes to resolve the walls of the wind tunnel as well as the on the wake behind the rotor. No mesh dependence study was performed.

The action of the rotor was modeled in the usual way by introducing body force-densities, see equations (1 - 3) along x, y and z directions respectively. These forces are regarded as to be originated from 2D airfoil data in combination with the BEM-technique. Velocities necessary for computing the total inflow velocity (eq. 4) are taken at each iteration loop. All can be done very easily using the UDF-interface provided by FLUENT.

$$mom_x = \rho W^2 Bc(C_l.\sin\phi - C_d.\cos\phi)\sin\theta/(4\pi rt) \quad (1)$$

$$mom_y = \rho W^2 Bc(C_l.\sin\phi - C_d.\cos\phi)\cos\theta/(4\pi rt) \quad (2)$$

$$mom_z = \rho W^2 Bc(C_l.\sin\phi + C_d.\cos\phi)/(4\pi rt) \quad (3)$$

$$V_{rel} = \sqrt{(r.\omega + (v.cos\theta - w.sin\theta))^2 + u^2}, \phi = tan^{-1}(u/V_{rel}) \quad (4)$$

In equations (1 - 4), ρ is air density, V_{rel} is relative velocity to the local airfoil, B is blade number, c is local chord, C_l and C_d are lift and drag coefficients of the airfoil, ϕ is flow angle, θ is azimuthal angle of the element in the disc plan, r is the local radius of the element, t is the rotor thickness, ω is angular velocity of the rotor

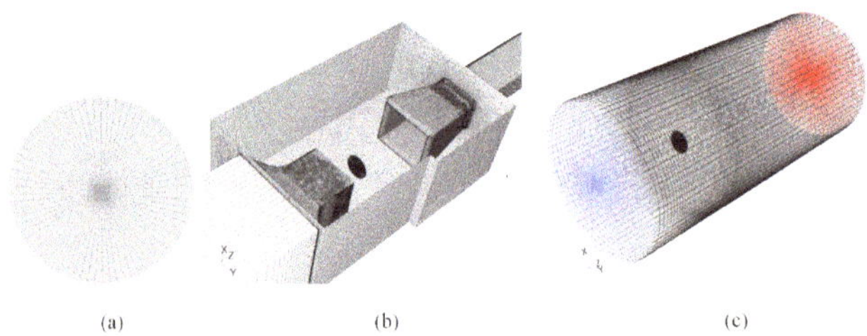

Fig. 1 (a) Structured mesh for the actuator disc, (b) Hybrid mesh for the LLF wind tunnel and (c) structured mesh for the boundary-free model

(44.45 $rads^{-1}$), u, v and w are axial (z-direction), vertical (x-direction) and horizontal (y-direction) components of the flow velocity on the element, respectively, x and y are vertical and horizontal distance of the element to the rotor center. In these equations, lift and drag coefficients are called from a airfoil data base. Flow velocity components (u, v, w) and the planar location components (x,y) of the element are measured from the Navier-Stocks domain in the iteration process. Equations (1 - 3) have been obtained from blade element analysis. Unknown variables in equation (1), (2) and (3) are calculated from equations (4) in the iteration process. Two different types of Subroutines were used within the UDF interface. The Source (body forces)-providing Subroutines and those for integrating thrust and power and their coefficients and averaged velocities in front of and behind the rotor. Both wind tunnel and boundary-free models were run under different inflow conditions.

3 Results and Discussion

The results of these runs are presented into four sections: Rotor Performance, blade loads, wake visualization and wake traverse lie-graphs.

3.1 Rotor Performance

Both full wind tunnel and boundary-free models show almost the same values for power and thrust. Power coefficients (Fig.2a) derived from our AD model exhibit remarkable agreement with experimental measurement (EX) for the 425 rpm case. The Thrust coefficient (Fig.2b) is overestimated compared to the measurement with a constant value in the order of 0.1. Results from the RANS model of [3] *DM* are also included. Whereas DM underestimates power constantly our (AD) results for the AD-thrust fits reasonable to DM nut give much larger values than measured.

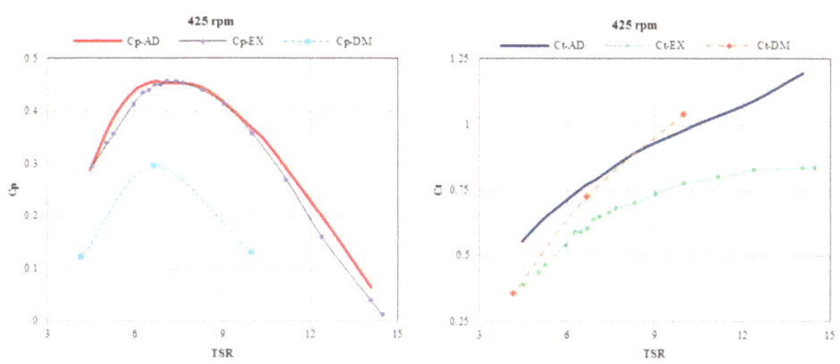

(a) (b)

Fig. 2 Power (a) and thrust (b) coefficients from AD and RANS (DM) models compared to experimental data

3.2 Radial Resolved Force Data

Both wind tunnel and boundary-free simulations of AD model result in approximately the same values for axial and tangential forces for the 425 RPM case (see fig. 3). Both AD and DM exceed the experimental measured values, especially when r/R > 0.6. The difference between AD and DM-Model is more pronounced for the tangential forces.

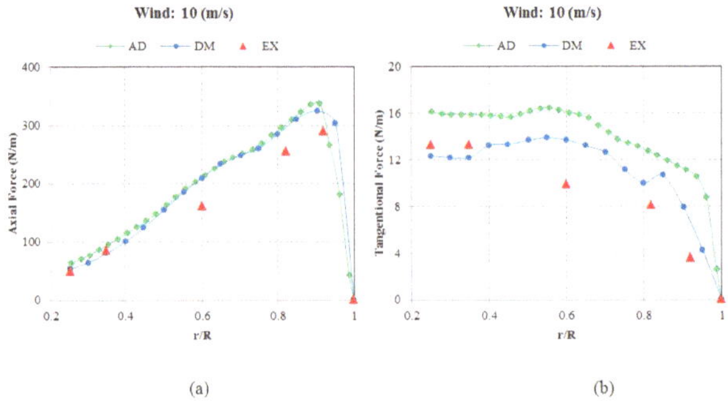

Fig. 3 Radial (spanwise) distribution of normal (a) and tangential (b) forces on one Mexico rotor blade

3.3 Wake Visualization

Fig. 4 shows the most important outcome of our work: the visualization of the near wake behind the rotor. It is clearly seen that the nozzle of the wind-tunnel is far too close to the disk (fig. 4-top), thereby preventing development of the free wake (fig. 4-down). The color scales are in the same range (from -5.11 $m.s^{-1}$ to 21.3 $m.s^{-1}$).

3.4 Near and Far Wake Traverse

Fig. 5 gives a more quantitative comparison of the development of the wake. Fig. (5a) shows axial-velocity in wind-direction whereas Fig. (5b) gives the same velocity in a rotor-plane about 1.4 rotor-diameters downstream. It can be seen that the differences between the tunnel-model and boundary-free model are smaller as compared to the measurements.

From Fig. (5a), is can be seen that wind tunnel model seems to have lower inlet velocity when compared to the boundary-free model.

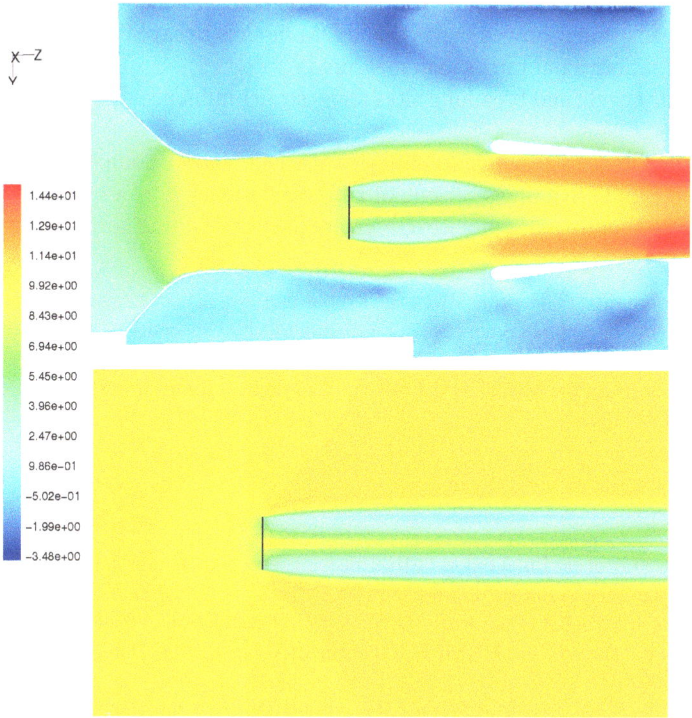

Fig. 4 Variation of axial velocity for the full wind tunnel model (top contour) and the boundary-free case (down contour) both in a range of -5.11 $m.s^{-1}$ to 21.3 $m.s^{-1}$

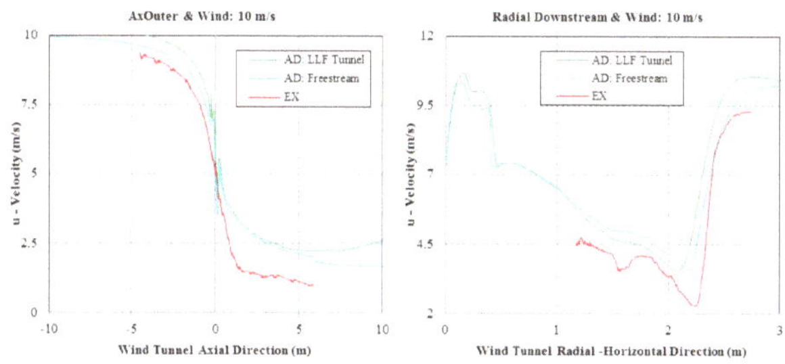

(a) (b)

Fig. 5 Axial velocity component along axial (a) and radial directions (b)

4 Summary and Conclusions

A differential Actuator Disk (AD) model was prepared using the commercial CFD-RANS code FLUENT 5.7 for the MEXICO rotor. The results were compared with a RANS model of Jeromin and Schaffarczyk and measurements. Our main conclusions are as follow:

- Power correlation for actuator disc exhibited good agreement with measurement.
- Blade loads study showed that AD and DM models behave the same for predicting much more axial forces than measured.
- Collector nozzle and breathing slots show strong influence to the wake development.
- Wake length decreased due to collusion the wind by the compressor nozzle and crossing from breathing slot zone of LLF wind tunnel.
- AD gives slightly more precise agreement than DM for the near wake modeling.

Acknowledgements. This project was carried out during a research stay of one of us (EM) at the University of Applied Sciences Kiel. Funding this by University of Tehran is gratefully acknowledged. Gerard Schepers from Energy research Center of Netherland (ECN) provides some hints for using the MEXICO database and Andreas Jeromin from UAS Kiel provided his RANS (DM) data.

References

1. Schepers, J.G., et al.: Final report of IEA Task 29, Mexnext (Phase 1): Analysis of Mexixo wind Tunnel measurements, ECN-E-12-004, Petten, The Netherlands (2012)
2. Micallef, D.: MEXICO Data Analysis, Stage IV - Investigation of the Limitations of BEM Codes on the Basis of the MEXICO Experiment: Version 2, TU Delft, ECN, The Netherlands (2009)
3. Jeromin, A., Schaffarczyk, A.P.: Full 3D computations of a rotating blade using unstructured grids. In: 3rd IEA Wind Annex 29 (MexNext) Meeting, Forth, Heraklion, Crete, Greece (2010)
4. Shen, W.Z., Zhu, W.J., Sϕrensen, J.N.: Actuator line/NavierStokes computations for the MEXICO rotor: comparison with detailed measurements. Wind Energy (2011), doi:10.1002/we.510
5. Breton, S.P., Watters, C.S., Masson, C.: Using the Actuator Surface Method to Model the Three-Bladed MEXICO Wind Turbine. In: Proceedings of the 48th AIAA Aerospace Sciences Meeting. AIAA (2010)
6. Rethore, P.E., Srensen, N.N., Madsen, H.A.: Modeling the MEXICO Wind Tunnel with CFD. In: 3rd IEA Wind Annex 29 (MexNext) Meeting, Forth, Heraklion, Crete, Greece (2010)

Thick Airfoil Deep Dynamic Stall

Hanns Mueller-Vahl, Christoph Strangfeld, Christian Navid Nayeri, Christian Oliver Paschereit, and David Greenblatt

1 Introduction

Recent interest in vertical axis wind turbines (VAWTs) has surged due to their appearance in the built environment and their potential for deep water offshore applications. Their well-known advantages include insensitivity to the wind direction, proximity of the generator to the ground and relatively low noise levels. Furthermore, blade profile uniformity along the span can significantly reduce manufacturing costs, particularly for large scale utility machines. A drawback of VAWTs is the tendency of their blades to stall dynamically when they are pitched beyond their static stall angle. Dynamic stall is characterized by a strong vortex, the dynamic stall vortex (DSV), which forms near the leading-edge. When the DSV is convected downstream, a rapid drop in lift and severe pitching moment fluctuations result [1]. On VAWTs operating at low tip-speed ratios, dynamic stall occurs periodically throughout the rotation of the blades [2]. This situation is unique in that the blades experience pitch oscillations about zero angle of attack and the flow separates alternately on both sides. This results in a number of undesired effects: On the one hand, the full separation on the suction surface produces a sharp drop in c_l and thus rotor torque. On the other hand, the unsteady aerodynamic loads cause fatigue damage to the generator and drive train [3].

Hanns Mueller-Vahl · David Greenblatt
Faculty of Mechanical Engineering, Technion - Israel Institute of Technology, Haifa 32000, Israel
e-mail: {hannsmv,davidg}@technion.ac.il

Christoph Strangfeld · Christian Navid Nayeri · Christian Oliver Paschereit
TU Berlin, ISTA, FG Experimentelle Stroemungsmechanik, Mueller-Breslau-Str. 8, D-10623 Berlin, Germany
e-mail: {strangfeld,christian.nayeri, oliver.paschereit}@tu-berlin.de

M. Hölling et al. (eds.), *Wind Energy – Impact of Turbulence*,
Research Topics in Wind Energy 2,
DOI: 10.1007/978-3-642-54696-9_6, © Springer-Verlag Berlin Heidelberg 2014

The vast majority of dynamic stall research to date was motivated by the load oscillations on helicopter blades and therefore focuses on thin airfoils at $Re > 10^6$ [1, 4]. In the built environment, however, VAWT blades experience Reynolds numbers that are approximately one order of magnitude lower. Moreover, thick blade sections that are employed for structural reasons exhibit a fundamentally different stall mechanism that is exacerbated at low Reynolds numbers [5]. The predominant use of thick blade sections at low Reynolds numbers, combined with the dearth of research [6], motivated an experiential study of VAWT dynamic stall and its control. This paper describes initial results of a study conducted at the Technion that is focused entirely on the dynamic stall mechanism under these unique conditions.

2 Experimental Setup and Procedure

Experiments were conducted in a blow-down wind tunnel driven by a 75kW radial blower. Wind tunnel turbulence and flow distortion in the upstream test section were less than 0.3% and 2% respectively. A schematic of the test section (inner dimensions: 610mm x 1004mm) that produces the rotation of the wing around the quarter-chord position is presented in Fig. 1. A Pitot static probe connected to a piezoresistive pressure transducer was used to monitor the instantaneous wind tunnel speed. The constant chord NACA 0018 airfoil model with a chord length of c=347mm and a span of s=610mm was machined from Obumodulan®, a synthetic material developed for model building. A cross-sectional view showing the two control slots at 5% and 50% chord is presented in Fig. 2. Even though the data presented here only includes cases without flow control, the forcing slots were not sealed to facilitate a direct comparison between control and baseline cases. The airfoil was equipped with 40 pressure ports (ϕ 0.8mm) distributed along the entire upper and lower surfaces. Two ESP-32HD piezoresistive pressure scanners were mounted inside the airfoil and connected to the pressure ports via vinyl tubing.

The measurements at Re=$2.5 \cdot 10^5$ were carried out at a wind tunnel speed of U_∞=11.1m/s and pitch rates of f=0.188Hz, 0.375Hz and 0.75Hz (corresponding to reduced frequencies of $k = \omega c/2U_\infty = 0.018$, 0.037 and 0.074 respectively). Some additional tests were conducted at Re=$5 \cdot 10^5$ (U_∞=22.2m/s). During the surface pressure measurements, pressure data was recorded for 85 seconds at a sample rate of 481.14Hz. Each measurement was repeated three times to ensure negligible statistical uncertainties. The lift coefficient c_l and moment coefficient c_m were numerically derived from phase averaged pressure distributions.

At selected phase angles, PIV measurements were carried out using a 200mJ double-pulsed Nd-YAG laser in conjunction with a 4 megapixel CCD camera. The PIV and pressure data acquisition were synchronized by a trigger output from the motor controller and 300 pitching cycles were recorded at each phase angle. A commercial seeding generator generated particles of diameter 1μm within the tunnel plenum. The laser head was mounted 850mm downstream of the trailing-edge of the airfoil. The light sheet was oriented perpendicular to the airfoil surface, parallel to the x and y-axes. A pulse separation of 40μs was selected. A decreasing size

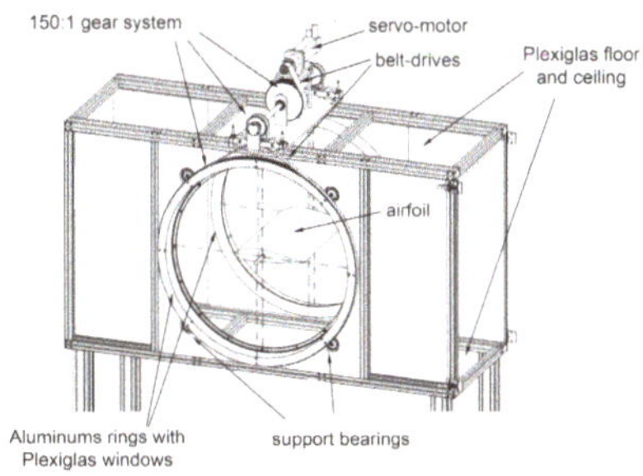

Fig. 1 View of the test section showing the servo-motor and belt drives as well as the approximate airfoil location

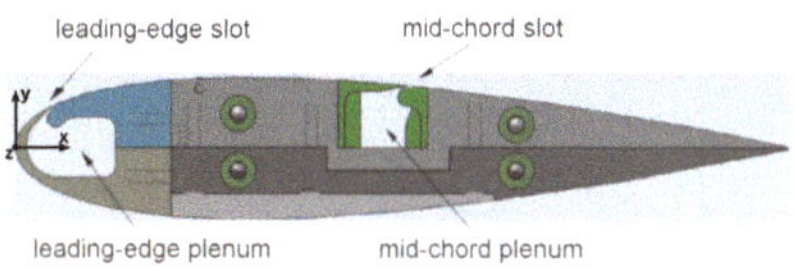

Fig. 2 Cross-sectional view of the NACA 0018 airfoil model showing the flow control slots

multi-pass correlation approach was adopted using the software DaVis by LaVision. The final results were calculated from 32x32 pixel interrogation windows with 50% overlap, producing a velocity field with a spacing of 3.2mm between two vectors. In the plots presented below, every third vector is shown in both directions for clarity. The normalized vorticity fields were computed according to $\omega_z = [\frac{\partial v}{\partial x} - \frac{\partial u}{\partial y}] \cdot \frac{c}{U_\infty}$.

3 Results

Data recorded at Re=2.5·10^5 for a pitching motion of α=0°+30°·sin(ωt) with a reduced frequency of k=0.074 is presented here in detail. In Figs. 3 and 4, plots of the static c_l and c_m, which are consistent with the results of Timmer [7], are shown along with unsteady data. In addition a series of simultaneously recorded phase averaged PIV and pressure data is shown in Fig. 5. The coordinate system is fixed with the wing and the instantaneous direction of the free stream is indicated by an arrow located at the axis of rotation. The pressure coefficient c_p on the upper surface is indicated by a dashed line and that on the lower surface by a dotted line. The markers (a) to (d) in Figs. 3 and 4 correspond to the velocity fields presented in Figure 5.

As the wing pitches up dynamically, the boundary layer across the rear half of the blade thickens. Around α=20°reversed flow caused by the adverse pressure gradient gradually progresses from the trailing-edge. This situation marks the beginning of moment stall (see Fig. 4). The drop in c_m is the result of the formation of a large vortex, termed the *aft dynamic stall vortex* (ADSV) here, across the rear half of the airfoil. Figure 5(c) illustrates that the presence of the ADSV near the airfoil surface results in a distinct region of low c_p. The surface pressure distribution prior to the formation of the ADSV (see Fig. 5(a)) is indicative of a separation bubble located at approximately $0.5 \leq x/c \leq 0.6$. This bubble is presumed to form into the ADSV. An analysis of the trajectory of the ADSV by means of the Weiss statement of the Q-value (see [8]), not shown here, confirmed this assumption. These observations should be contrasted with common descriptions of thin airfoil dynamic stall (see

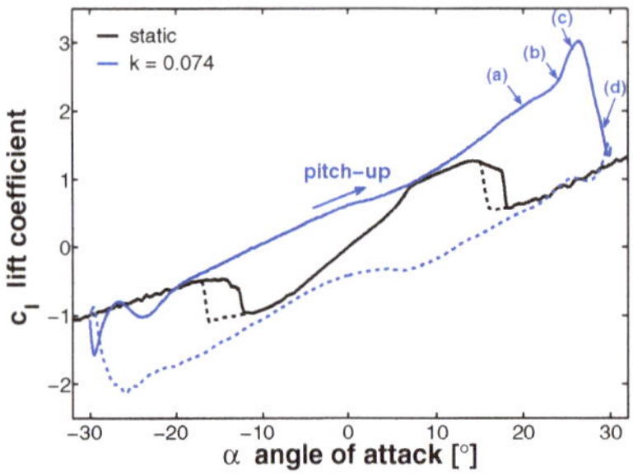

Fig. 3 Comparison of the static lift coefficient to dynamic pitching with $\alpha=0°+30°\cdot\sin(\omega t)$, k=0.074, $Re=2.5\cdot10^5$

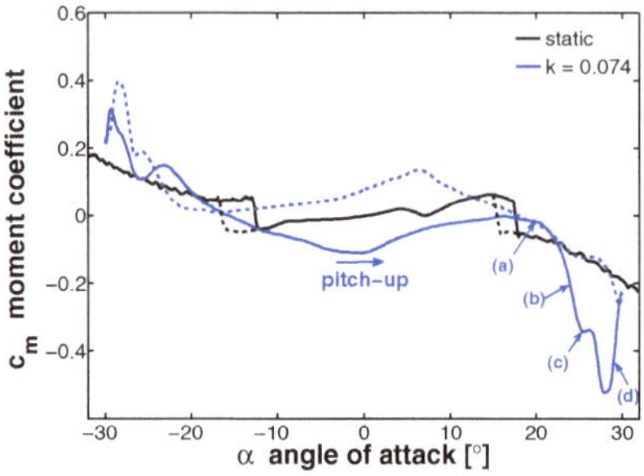

Fig. 4 Comparison of the static moment coefficient to dynamic pitching with $\alpha=0°+30°\cdot\sin(\omega t)$, k=0.074, $Re=2.5\cdot10^5$

references in [1]), where the moment remains almost constant until the shedding of the leading-edge DSV (LEDSV) results in sharp fluctuations: The drop in c_m observed here occurs before the LEDSV has even reached a considerable size. At $Re=5\cdot10^5$ (data not shown here), the ADSV is weaker for a given k.

Starting at approximately $\alpha=24°$, the growth of the DSVs produces an increase in the lift slope. At $\alpha=25.5°$, two distinct low pressure regions on the upper surface corresponding to the positions of the DSVs are clearly visible (see Fig. 5(c)), which can be interpreted as a result of the high induced velocity near the wing surface. The subsequent shedding of the LEDSV and the associated motion of the corresponding low pressure region downstream lead to the second moment coefficient dip. Simultaneously, the ADSV lifts off the airfoil surface, remains in place and weakens steadily before merging with the LEDSV. Following the passage of the leading-edge DSV, the shear layer is fully separated, causing further loss in lift and moment recovery. As the wing pitches back down, the shear layer remains separated even below the static stall angle of $\alpha_s=18°$. The flow then reattaches front to back.

A comparison of c_l and c_m for various reduced frequencies is presented in Figs. 6 and 7. While for k=0.018, the lift coefficient almost attains static values at $\alpha=0°$, strong dynamic effects are visible throughout the entire pitching cycle at k=0.074: The lift zero crossings are at $\alpha=-10.8°$and $\alpha=12.2°$. With increasing k, the development of the LEDSV and its convection are delayed to higher angles of attack, which is consistent with results for rotorcraft blades [1]. Its shedding starts at $\alpha\approx19.5°$(k=0.018), $\alpha\approx21°$(k=0.037) and $\alpha\approx24°$(k=0.074) respectively. Correspondingly, higher values of the suction peak are attained. At higher pitch rates, the LEDSV attains a larger strength. These differences are reflected in the fact that for higher values of k, lift and moment stall are more severe and occur at a later stage of the pitching cycle. The strength of the ADSV is also highly dependent on the reduced frequency: While it has a large impact on the surface pressure at k=0.074, it is less pronounced at k=0.037 and hardly detectable at k=0.018. Accordingly, the initial moment dip resulting from the low pressure region caused by the ADSV is far less strong at k=0.018. This may in part be explained by the fact that the formation

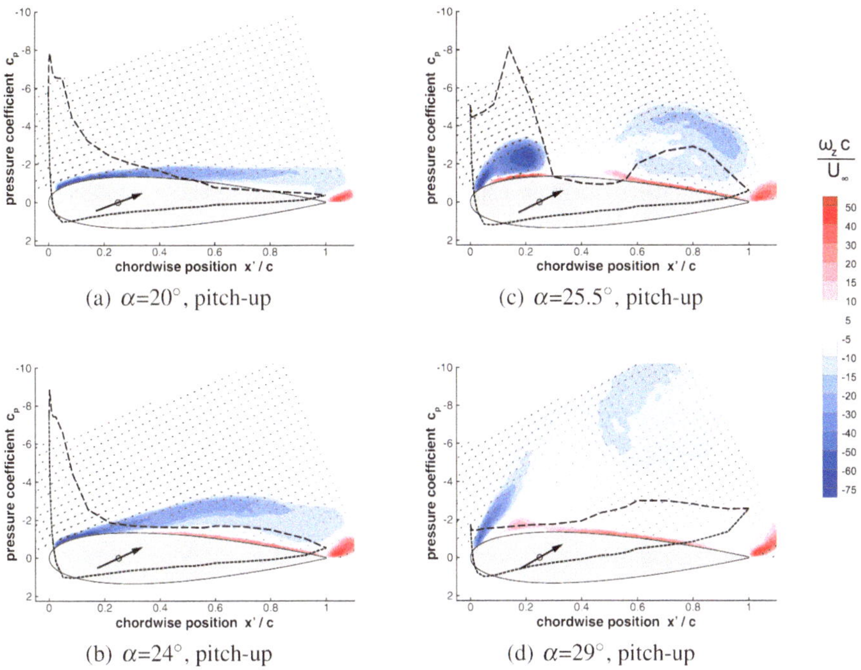

Fig. 5 Phase averaged normalized vorticity fields and surface pressure coefficients illustrating various stages of the dynamic stall mechanism, $\alpha=0°+30°\cdot\sin(\omega t)$, k=0.074, $Re=2.5\cdot10^5$

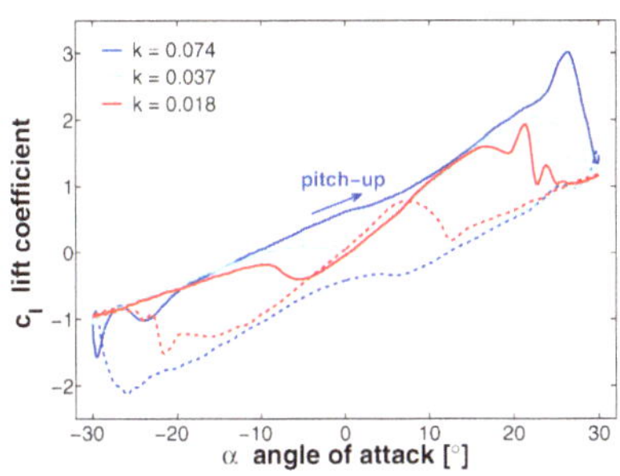

Fig. 6 Phase averaged unsteady lift coefficient for various reduced frequencies, $\alpha=0°+30°\cdot\sin(\omega t)$, $Re=2.5\cdot10^5$

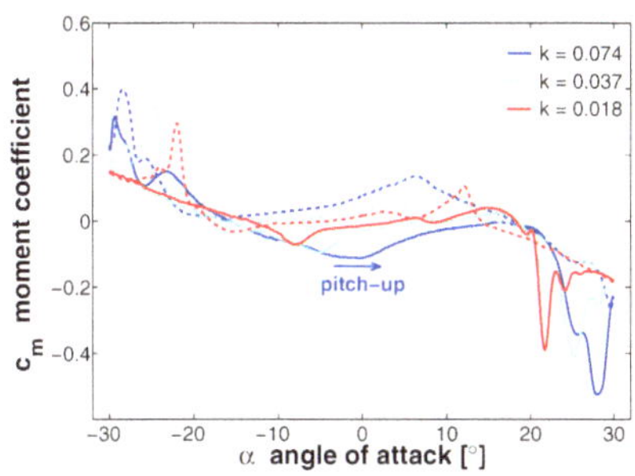

Fig. 7 Phase averaged unsteady moment coefficient for various reduced frequencies, $\alpha=0°+30°\cdot\sin(\omega t)$, $Re=2.5\cdot10^5$

of the ADSV always commences at approximately $\alpha\approx20°$; the earlier shedding of the LEDSV at lower pitching frequencies may disrupt its further growth.

The asymmetry of the c_l and c_m curves around α=0° is quite significant given that a symmetric airfoil model is investigated here. This asymmetry may be caused by the presence of the flow control slots. However, as the pressure distributions on the smooth airfoil surface (not shown) also display a low pressure region indicative of an ADSV, the control slots do not appear to play a major role in its formation.

4 Conclusions

The dynamic stall mechanism on a NACA 0018 airfoil observed here exhibits fundamental differences to the established findings regarding dynamic stall on rotorcraft blades. An additional dynamic stall vortex, never described before, was observed to form across the rear half of the blade. At a reduced pitching frequency of k=0.074, it produces a significant drop in the moment coefficient prior to the shedding of the leading-edge dynamic stall vortex. At lower values of k, the aft dynamic stall vortex is weaker. Increasing the Reynolds number also reduces its impact. It is presumed here that this phenomenon is the combined result of the airfoil thickness and the low Reynolds number.

References

1. Carr, L.W.: Progress in analysis and prediction of dynamic stall. Journal of Aircraft 25(1), 6–17 (1988)
2. Ferreira, C.S., Kuik, G.V., Bussel, G.V., Scarano, F.: Visualization by piv of dynamic stall on a vertical axis wind turbine. Experiments in Fluids 46(1), 97–108 (2009)
3. Paraschivoiu, I.: Double-multiple streamtube model for darrieus wind turbines. In: Second DOE/NASA Wind Turbines Dynamics Workshop, NASA CP-2186, Cleveland, OH, pp. 19–25 (February 1981)
4. Mulleners, K., Raffel, M.: The onset of dynamic stall revisited. Exp. Fluids (2012) 52, 779–793 (2012)
5. Greenblatt, D., Wygnanski, I.: Effect of leading-edge curvature on airfoil separation control. Journal of Aircraft 40(3), 473–481 (2003)
6. Wickens, R.H.: Wing tunnel investigation of dynamic stall of an naca 0018 airfoil oscillating in pitch. Tech. rep (1985)
7. Timmer, W.A.: Two-dimensional low-reynolds number wind tunnel results for airfoil naca 0018. Wind Engineering 32(6), 525–537 (2008)
8. Carmer, C.F.: Evaluation of Large-Scale Wing Vortex Wakes from Multi-Camera PIV Measurements in Free-Flight Laboratory. In: Particle Image Velocimetry: New Developments and Recent Applications, vol. 1, pp. 377–394. Springer (2008)

Part II
Turbulence

The Scaling Properties of the Turbulent Wind Using Empirical Mode Decomposition and Arbitrary Order Hilbert Spectral Analysis

Rudy Calif, François G. Schmitt, and Yongxiang Huang

Abstract. In this work, we present an analysis of one-year period measured wind speed in the atmospheric boundary layer from a wind energy production site. We employ a Hilbert-based methodology, namely arbitrary-order Hilbert spectral analysis to characterize the intermittent property of the wind speed in a joint amplitude-frequency space. The measured scaling exponents implies intermittent nature of the wind on mesoscales.

1 Introduction

Turbulent atmospheric wind speed is a complex process with a very large Reynolds number *Re* [1]. This Reynolds number leads to huge intermittency of wind speed fluctuations involving a wide range of temporal and spatial scales (the planet scale to the dissipative scale) [2]. Knowledge of the dynamics of this process is crucial for wind energy applications. Several works have highlighted the universality of the scaling and intermittent properties of turbulent wind speed in the inertial range in

Rudy Calif
EA, LARGE Laboratoire en Géosciences et énergétique,
Université des Antilles et de la Guyane 97170 P-á-P
e-mail: rudy.calif@univ-ag.fr

François G. Schmitt
CNRS, UMR 8187 LOG Laboratoire d'Océanologie et de Géosciences,
Université de Lille 1, 28 Avenue Foch, 62930 Wimereux, France
e-mail: francois.schmitt@univ-lille1.fr

Yongxiang Huang
Shanghai Institute of Applied Mathematics and Mechanics,
Shanghai Key Laboratory of Mechanics in Energy Engineering, Shanghai University,
Shanghai 200072, People's Republic of China
e-mail: yongxianghuang@gmail.com

M. Hölling et al. (eds.), *Wind Energy – Impact of Turbulence*,
Research Topics in Wind Energy 2,
DOI: 10.1007/978-3-642-54696-9_7,

the atmospheric boundary layer [3, 4, 5, 6]. However the knowledge of variations ranging from minutes to few days corresponding to 1 to 1000 km, i.e. the mesoscale range, is necessary to provide efficient information for management and control of the wind power generation. The studies concerning this scale range are less numerous than that for the small-scale range, due to the possible non-universality of the power law slope in the mesoscale range [7, 8]. Recent works [7, 8] have been dedicated to scaling and multiscaling properties of the atmospheric wind speed in the mesoscale range. They highlighted the multiscaling and intermittency properties of atmospheric surface layer-winds in the mesoscale range. However, as pointed by Huang et al. [18, 20] that the traditional methodologies might be strongly influenced by large-scale energetic structures, e.g, very-large-scale-motion in atmospheric boundary layer [21]. In this paper, the scaling properties of the atmosphere are investigated using a new Hilbert-based approach, namely arbitrary-order Hilbert spectral analysis, in which the large-scale influence can be constrained [9, 18, 20].

2 Arbitrary-Order Hilbert Spectral Analysis

The arbitrary-order Hilbert spectral analysis [9, 10] is an extended version of Hilbert-Huang Transform (HHT) [11, 12], which is designed to characterize the scale invariance of a given time series in a joint amplitude-frequency space. The same as the HHT, it proceeds in two steps: i) empirical mode decomposition, ii) Hilbert spectral analysis. Empirical mode decomposition is an efficient tool to separate a nonlinear and nonstationary time series into a sum of Intrinsic Mode Functions without *a priori* basis as required by traditional Fourier-based method [11, 12, 13]. An IMF must respect two conditions: i) the difference between the number of local extrema and the number of zero crossings must be zero or at most one, ii) the running mean value of two envelopes estimated by the local maxima and local minima is zero [11, 12]. Thus the original signal $u(t)$ is decomposed in a sum of $n-1$ IMF modes with the residual $r_n(t)$

$$u(t) = \sum_{m=1}^{n-1} C_m(t) + r_n(t) \tag{1}$$

To obtain a physical meaningful IMF, this sifting process must be stopped by a certain criterion [11, 12]. More details of the EMD decomposition are given in Refs. [11, 12, 14, 13, 17].

In order to determine the time-frequency energy distribution from the original signal $P(t)$, HSA is performed to each obtained IMF component $C_m(t)$ to extract the instantaneous amplitude $\mathscr{A}_m(t)$ and frequency $\omega_m(t)$ [11, 15, 16]. The Hilbert transform is written as

$$\tilde{C}_m(t) = \frac{1}{\pi} U \int_{-\infty}^{+\infty} \frac{C_m(s)}{t-s} ds \tag{2}$$

with U the Cauchy principle value [15, 16]. One can define an analytical signal $z_m(t)$ for each IMF mode $C_m(t)$, i.e.,

$$z_m(t) = C_m(t) + j\tilde{C}_m(t) = \mathscr{A}_m(t)e^{j\varphi_m(t)} \tag{3}$$

where $\mathscr{A}_m(t) = |z_m(t)| = \sqrt{C_m(t)^2 + \tilde{C}_m(t)^2}$ represents an amplitude and $\varphi_m(t) = \arg(z) = \arctan\left[\frac{\tilde{C}_m(t)}{C_m(t)}\right]$ represents the phase function of IMF modes. Hence, the instantaneous frequency $\omega_m(t)$ is defined from the phase $\varphi_m(t)$

$$\omega_m(t) = \frac{1}{2\pi}\frac{d\varphi_m(t)}{dt} \tag{4}$$

The original signal $P(t)$ can be expressed as

$$P(t) = \mathscr{R}e\sum_{m=1}^{N}\mathscr{A}_m(t)e^{j\varphi_m(t)} = \mathscr{R}e\sum_{m=1}^{N}\mathscr{A}_m(t)e^{j\int_{-\infty}^{t}\omega_m(t)dt} \tag{5}$$

in which $\mathscr{R}e$ is part real [11, 12, 9].

In order to characterize the scale invariant property of a considered signal in the Hilbert frame, Huang et al. [10] proposed an extension of HHT, arbitrary-order Hilbert spectral analysis. This approach has been applied successfully on turbulence data [10, 18], river discharge data [19]. For that, a joint pdf $p(\mathscr{A},\omega)$ of the instantaneous frequency ω and amplitude $\mathscr{A}$ from all these IMF components is extracted [10, 9]. In this frame, the Hilbert marginal spectrum is rewritten as

$$h(\omega) = \int_0^{+\infty} p(\omega,\mathscr{A})\mathscr{A}^2 d\mathscr{A} \tag{6}$$

This expression concerns only the second-order statistical moment. A generalization of this definition is considered to arbitrary-order statistical moment $q \geq 0$ [10, 9, 20]

$$\Psi_q(\omega) = \int_0^{+\infty} p(\omega,\mathscr{A})\mathscr{A}^q d\mathscr{A} \tag{7}$$

Hence, in the Hilbert space, the scale invariance is written as $\Psi_q(\omega) \sim \omega^{-\xi(q)}$, where $\xi(q)$ is the corresponding scaling exponent in the Hilbert space. This scaling exponent function can be linked to scaling exponent function $\zeta(q)$ of structure functions by $\zeta(q) = \xi(q) - 1$ [10, 9, 20]. Here the Hurst exponent could be defined as $H = \xi(1) - 1$ [20].

3 Results

We present the analysis of the measured atmospheric wind speed wtih a sampling rate of 1 Hz, obtained from a wind energy production site of Petit Canal in Guadeloupe, an island in the West Indies, located at 16°15N latitude 60°30 W longitude. Figure 1 illustrates the power spectral density provided by the Fourier transform and the Hilbert transform. The corresponding scaling exponents are $\beta = 1.27$ and $\xi(2) = 1.28$ respectively in Fourier and Hilbert spaces, over a frequency range

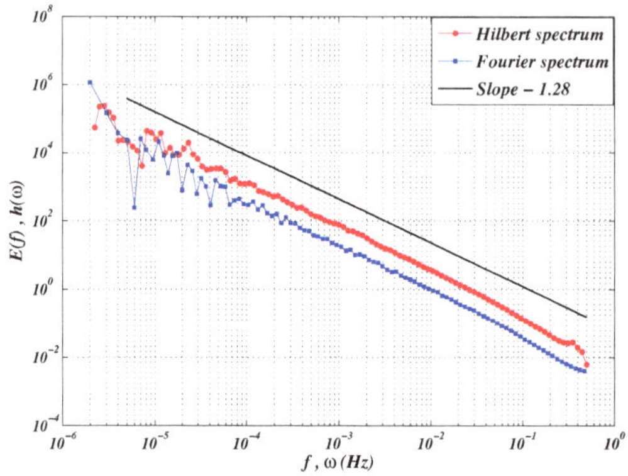

Fig. 1 Power spectral densities for wind speed $u(t)$ provided by the Fourier transform, $E(f)$ (□) and the Hilbert approach, $h(\omega)$ (○). Power law behavior is observed on a large range of scales with a scaling exponent 1.28.

$10^{-5} \leqslant f \leqslant 0.5$ Hz. If one applies the Taylor's frozen hypothesis [22], this corresponds to a spatial scale from 16 m to 800 km. Figure 2 shows the measured mean frequency f_m of each IMF mode. Here we estimate the mean frequency by the following definition [11, 19]:

$$f_m = \frac{\int_0^\infty f E_m(f) df}{\int_0^\infty E_m(f) df} \tag{8}$$

where $E_m(f)$ is the Fourier spectrum of mth IMF mode C_m. It is an energy weighted mean frequency in the Fourier space [11, 19]. We obtain a relation of the form

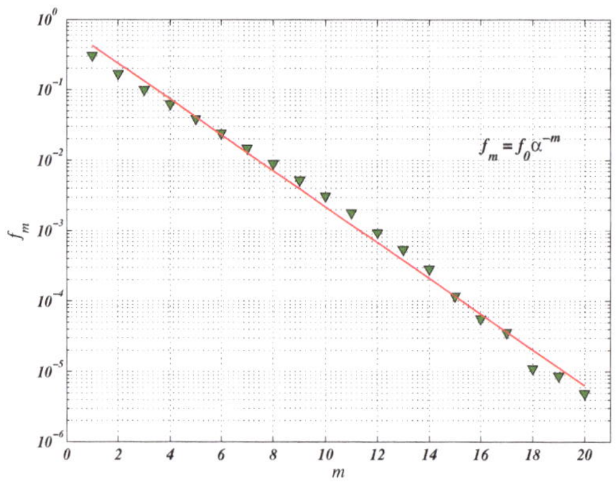

Fig. 2 Representation of the mean frequency f_m versus the mode index m, in log-linear plot. The fitting slope is 0.58, corresponding to $\alpha = 1.8$-times filter bank.

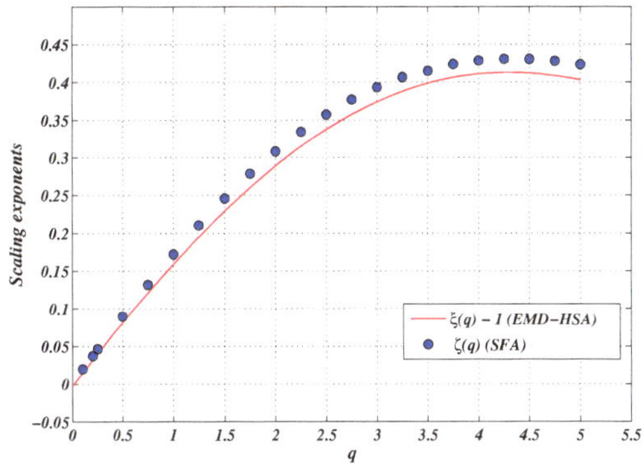

Fig. 3 Scaling exponents measured by the structure function analysis (SFA) and arbitrary-order Hilbert spectral analysis (EMD-HSA)

$f_m = f_0\alpha^{-m}$, with $f_0 = 1.34$ and $\alpha = 1.8$, meaning that f_m of one IMF is approximately 1.8-times f_m of the next one. This characteristic corresponds to an almost dyadic filter bank in the frequency domain. This filter bank property is deduced by the data set itself, showing the fully adaptiveness of the method [13]. Figure 3 displays the corresponding scaling exponents $\xi(q) - 1$ (red line) obtained from arbitrary-order Hilbert spectral analysis and $\zeta(q)$ ($\circ$) obtained from classical structure function analysis. We can see that the scaling exponents $\xi(q) - 1$ are close to $\zeta(q)$: the obtained curves are concave and nonlinear, validating the intermittent and multiscaling properties of wind speed $u(t)$ in mesoscales.

4 Conclusion and Discussion

In this paper, we apply arbitrary-order Hilbert spectral analysis, to the measured wind speed of atmospheric boundary layer. We observe a power law behavior on a large range of time scales from $10^{-5} < f < 0.5\,\text{Hz}$. This can be associated to the spatial scales from 16 m to 800 km if one applies the Taylor's frozen hypothesis. It is found that the EMD decomposition acts as a filter bank, which is close to the dyadic one reported by several authors, showing the adaptiveness of the method. Furthermore, the intermittent nature of the wind speed is characterized by the scaling exponents $\zeta(q)$ and $\xi(q)$ respectively provided the classical structure function analysis and the Hilbert-based method. Despite the use of different methodologies (resp. different analysis frameworks), the coincidence of two scaling curves confirms the intermittent nature of the flow in the atmospheric boundary layer. Therefore, this intermittency must be taken into account when a model is proposed.

Acknowledgements. This study is financially supported by the Regional Council of Guadeloupe, European funding ($N^O 1/1.4/-31614$). Y. H. is financed in part by the National Natural Science Foundation of China under Grant No. 11072139.

References

1. Burton, T., Sharpe, D., Jenkins, N., Bossanyi, E.: Wind energy: handbook. Wiley Online Library (2001)
2. Lovejoy, S., Schertzer, D.: Towards a new synthesis for atmospheric dynamics: Space-time cascades. Atmos. Res. 96, 1–52 (2010)
3. Schmitt, F., Schertzer, D., Lovejoy, S., Brunet, Y.: Estimation of universal multifractal indices for atmospheric turbulent velocity fields. Fractals 1(3), 568–575 (1993)
4. Katul, G.G., Chu, C.R., Parlange, M.B., Albertson, J.D., Ortenburger, T.A.: Low-wavenumber spectral characteristics of speed and temperature in the atmospheric surface layer. J. Geoph. Res. 100(D7), 14,243-14,255 (1995)
5. Schmitt, F.G.: Gusts in intermittent wind turbulence and the dynamics of their return times. In: Peinke, J., Schaumann, P., Barth, S. (eds.) Proceedings of the Euromech Colloquium Wind Energy, pp. 73–79. Springer (2007)
6. Böttcher, F., Barth, S., Peinke, J.: Small and large fluctuations in atmospheric wind speeds. Stoch. Res. Ris. Asses. 21, 299–308 (2007)
7. Lauren, M.K., Menabe, M., Seed, A.W., Austin, G.L.: Characterisation and simulation of the multiscaling properties of the energy-containing scales of horizontal surface layer winds. Boundary Layer Meteorology 90, 21–46 (1999)
8. Muzy, J.F., Baile, R., Poggi, P.: Intermittency of surface-layer wind velocity series in the mesoscale range. Phys. Rev. E 81, 056308 (2010)
9. Huang, Y.: Arbitrary-order hilbert spectral analysis: Definition and application to fully developed turbulence and environmental time series. Ph.D. thesis, Université des Sciences et Technologies de Lille - Lille 1, France & Shanghai University, China (2009)
10. Huang, Y., Schmitt, F., Lu, Z., Liu, Y.: An amplitude-frequency study of turbulent scaling intermittency using hilbert spectral analysis. Europhys. Lett. 84, 40010 (2008)
11. Huang, N.E., Shen, Z., Long, S.R., Wu, M.C., Shih, H.H., Zheng, Q., Yen, N., Tung, C.C., Liu, H.H.: The empirical mode decomposition and the Hilbert spectrum for nonlinear and non-stationary time series analysis. Proc. R. Soc. London, Ser. A 454(1971), 903–995 (1998)
12. Huang, N.E., Shen, Z., Long, S.R.: A new view of nonlinear water waves: The Hilbert Spectrum. Annu. Rev. Fluid Mech. 31(1), 417–457 (1999)
13. Flandrin, P., Gonçalvès, P.: Empirical Mode Decompositions as Data-Driven Wavelet-Like Expansions. Int. J. Wavelets, Multires. Info. Proc. 2(4), 477–496 (2004)
14. Huang, N.E., Wu, M.L., Long, S.R., Shen, S.S.P., Qu, W., Gloersen, P., Fan, K.L.: A confidence limit for the empirical mode decomposition and Hilbert spectral analysis. Proc. R. Soc. London, Ser. A 459(2003), 2317–2345 (2037)
15. Cohen, L.: Time-frequency analysis. Prentice Hall PTR, Englewood Cliffs (1995)
16. Long, S.R., Huang, N.E., Tung, C.C., Wu, M.L., Lin, R.Q., Mollo-Christensen, E., Yuan, Y.: The Hilbert techniques: an alternate approach for non-steady time series analysis. IEEE Geoscience and Remote Sensing Soc. Lett. 3, 6–11 (1995)
17. Rilling, G., Flandrin, P.: One or two frequencies? The empirical mode decomposition answers. IEEE Trans. Signal Process, 85–95 (2008)
18. Huang, Y., Schmitt, F., Lu, Z., Fougairolles, P., Gagne, Y., Liu, Y.: Second-order structure function in fully developed turbulence. Phys. Rev. E 82, 26319 (2010)

19. Huang, Y., Schmitt, F.G., Lu, Z., Liu, Y.: Analysis of daily river flow fluctuations using empirical mode decomposition and arbitrary order hilbert spectral analysis. J. Hydrol. 373, 103–111 (2009)
20. Huang, Y., Schmitt, F.G., Hermand, J.-P., Gagne, Y., Lu, Z., Liu, Y.: Arbitrary-order Hilbert spectral analysis for time series possessing scaling statistics: comparison study with detrended fluctuation analysis and wavelet leaders. Phys. Rev. E 84, 016208 (2011)
21. Marusic, I., Mathis, R., Hutchins, N.: Predictive model for wall-bounded turbulent flow. Science, 193–196 (2010)
22. Frisch, U.: Turbulence: the legacy of AN Kolmogorov. Cambridge University Press (1995)

Multifractal Statistical Methods and Space-Time Scaling Laws for Turbulent Winds

G. Fitton, I. Tchiguirinskaia, D. Schertzer, and S. Lovejoy

Abstract. We discuss the results of a universal multifractal (UM) analysis performed on the GROWIAN wind speed dataset. Within this framework the wind speed can be reproduced, including the extremes, at all scales using just three parameters: α, C_1 and H [1]. We exploit the fact that the wind speed is simultaneously recorded at several positions (effectively two grids) on two masts. The first grid allows us to compare the scaling of the horizontal spatial increments of the wind speed (at three heights) with that of the temporal increments, thus enabling us to verify Taylor's hypothesis of frozen turbulence. The second grid allows us to test the hypothesis of scaling anisotropy between horizontal and vertical shears of the wind speed. The two scaling laws refer to the choice of either Kolmogorov energy or buoyancy force fluxes. The spatial structure function analyses assume the large number of data samples (approximately 150 samples of twenty minutes) reduces the uncertainties from the limited number of spatial points. The proof of universal scaling behaviour for different wind farm sites (see [2] for comparison) is an exciting concept that opens up the possibility of further areas of research and application within the field.

Data. Courtesy of Dr. Peinke we were introduced to the GROWIAN dataset. The experiment consisted of two 150m masts positioned 65m East-South-East of a 3MW wind turbine with a 52m lateral distance between each mast. Installed on the masts were 20 propellor anemometers positioned in pairs on the ends of booms with a 12m length. This covered an effective area of 75×100m. Wind speed, direction and temperature were measured at 2.5Hz and the duration of one measuring run was approximately 20 minutes. A total of 300 runs were sampled between April 1984 and February 1987 at different inflow conditions, however, errors in the data meant

G. Fitton · I. Tchiguirinskaia · D. Schertzer
Universitè Paris Est, Ecole des Ponts ParisTech, LEESU, 6-8 avenue B. Pascal,
Citè Descartes, 77455, Marne-la-Vallèee, France
e-mail: fittong@cereve.enpc.fr

S. Lovejoy
McGill University, Physics Department, 3600 University Street, Montreal, Quebec, Canada

M. Hölling et al. (eds.), *Wind Energy – Impact of Turbulence*,
Research Topics in Wind Energy 2,
DOI: 10.1007/978-3-642-54696-9_8, © Springer-Verlag Berlin Heidelberg 2014

the number of usable samples varied between 150 to 290. For more details about the experiment see [3].

Spectra. The field scales when the logarithm of the spectral energy, $E(\omega) \propto \omega^{\beta}$, has linear behaviour with a unique slope, $\beta = 2H + 1 - K(2)$ (see Fig. 1). The second order scaling moment function, $K(2)$, is an intermittency correction. The Hurst exponent, H, quantifies the divergence from conservation of a flux and is 1/3 for Kolmogorov energy and 3/5 for Bolgiano-Obukhov buoyancy fluxes. For data analysis, the horizontal wind speed, u, is represented as the two-dimensional vector, $\mathbf{u} = (u_x, u_y)$, preselected to be approximately perpendicular to the measuring array.

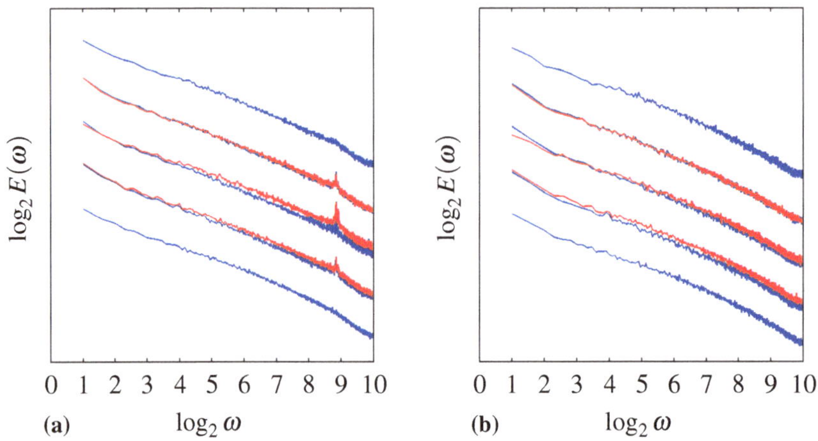

Fig. 1 Log-log plot of averaged wind speed spectra, $E(\omega)$ for all samples (a) and $E_{\perp}(\omega)$ for samples with direction approximately perpendicular to the array (b). The frequency, $\omega = 2^N/\ell_n = \lambda/0.4\text{s}$, where $\ell_n = 2^n \times 0.4\text{s}$ for $n \in [0 : N]$. Plots (shifted) from bottom to top correspond to wind speeds at 50, 75, 100, 125 and 150m measured on Mast 2. The blue and red correspond to inner and outer positions on the masts.

Wind Speed Direction. In Fig. 1a a spectral spike can be seen occurring at approximately $\log_2 \omega = 9$ or two seconds. In [4], a filtering criteria based on mean interpolated wind speed and turbulence intensity was used to supposedly remove measurements subject to a 'shadow effect' from the mast. The only way we could obtain the desired scaling (Fig. 1b) was to filter the wind speed so that it was perpendicular to the mast. This confirms the effect of the wakes of the anemometers causing the over representation of the two second frequency in Fig. 1a. In fact, we can further confirm this idea by estimating a characteristic length scale of 30m, given a mean wind speed of 10m/s, which is approximately the length of the boom.

UM Parameters. After removing the 'problematic' data we estimated α and C_1 – the UM parameters that measure respectively the multifractality and mean intermittency of the scaling field – over two subranges: $\log_2 \omega \in [1:7]$ & [8:10] using the DTM method [1]. We use the DTM method [1] on the third power of the time-series of the wind speed increments, $\Delta u = \sqrt{[\Delta u_x^2 + \Delta u_y^2]}$, and component increments, Δu_x

and Δu_y, for α and C_1, and the first-order structure function for H (see Table 1 for the average estimates).

Table 1 Estimation of UM parameters over *high* (top) and *low* (bottom) frequency subranges

Height	$\alpha_{\Delta u}$	$C_{1,\Delta u}$	$H_{\Delta u}$	$\alpha_{\Delta u_x}$	$C_{1,\Delta u_x}$	$H_{\Delta u_x}$	$\alpha_{\Delta u_y}$	$C_{1,\Delta u_y}$	$H_{\Delta u_y}$
150m	1.42	0.31	0.42	1.36	0.36	0.38	1.36	0.36	0.45
125m	1.40	0.32	0.37	1.36	0.37	0.33	1.35	0.36	0.42
100m	1.38	0.30	0.45	1.33	0.35	0.42	1.32	0.35	0.49
75m	1.39	0.30	0.45	1.35	0.36	0.42	1.34	0.36	0.49
50m	1.43	0.32	0.48	1.38	0.38	0.44	1.38	0.38	0.51
10m	1.50	0.34	0.47	1.47	0.43	0.43	1.47	0.42	0.51
150m	1.76	0.26	0.33	1.69	0.42	0.30	1.69	0.40	0.36
125m	1.71	0.26	0.30	1.64	0.37	0.26	1.64	0.37	0.34
100m	1.71	0.24	0.33	1.66	0.37	0.30	1.63	0.35	0.36
75m	1.66	0.24	0.31	1.59	0.37	0.30	1.54	0.31	0.34
50m	1.63	0.22	0.30	1.58	0.34	0.28	1.57	0.31	0.33
10m	1.67	0.20	0.26	1.61	0.33	0.23	1.61	0.30	0.29

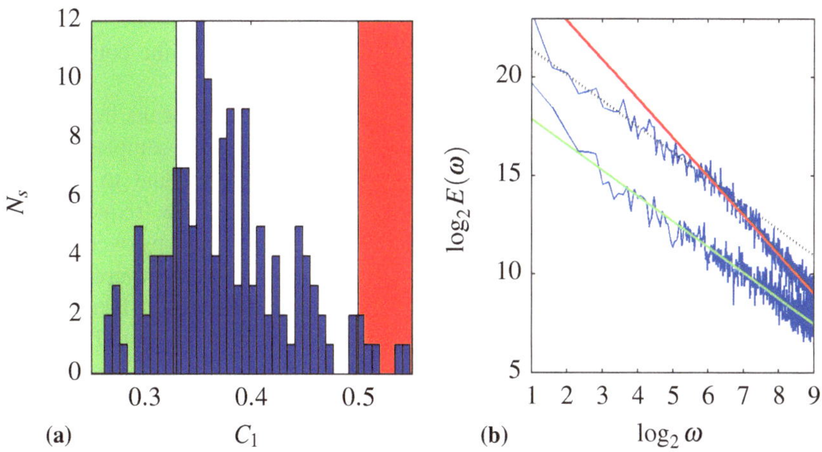

Fig. 2 Histogram of the parameter H for u_y at 75m (a), estimated from the first order structure function, using 152 subsamples, N_s. The estimates are comparable to those of Table 1. Summing the subsamples for extreme values $H < 0.33$ (green area) and $H > 0.5$ (red area) we get the corresponding spectral slopes (b) 2 (red), 1.4 (green) and 1.3 (dotted).

The estimates of H, show considerable variability, fluctuating around 0.3 and 0.6 i.e. between Kolmogorov and Bolgiano-Obukhov scaling (Fig. 2a). This is further supported by average spectra (Fig. 2b) selected so that they correspond to the extreme values of H. The mean codimension, C_1, fluctuates in a corresponding manner between 0.35 and 0.45. The multifractality, $\alpha \approx 1.5$, remains fairly stable at all heights and between horizontal components. The values are comparable with those found in [2]. To check the high frequency changes in scaling are not due to

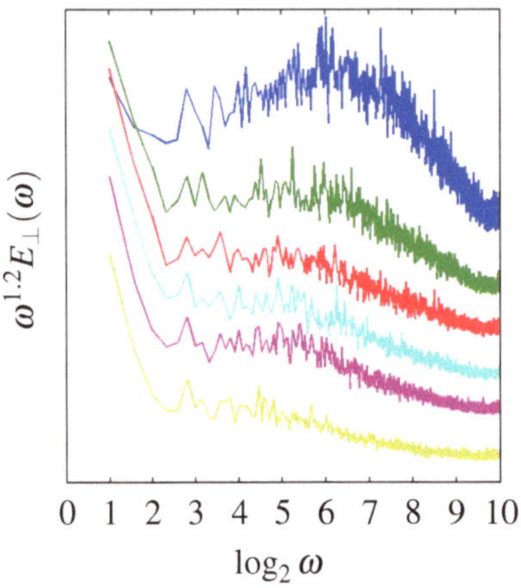

Fig. 3 Log-linear plot of *compensated* perpendicular wind spectra, $\omega^{1.2}E_\perp(\omega)$, vs. the normalised frequency, ω; Wind speeds at 10, 50, 75, 100, 125 and 150m, on the inner position of Mast 2 (shifted from top to bottom)

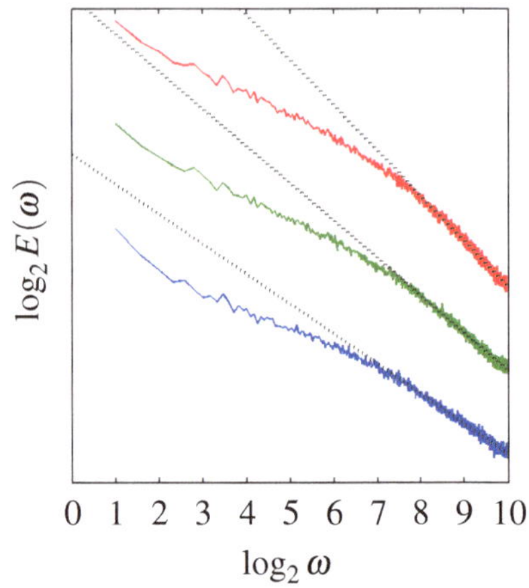

Fig. 4 Comparison of high frequency spectral slopes at 10m for u (blue), u_x (green) and u_y (red), shifted for easier comparison. The corresponding slopes of the dotted lines of best fit are 2.3, 1.9 and 1.5.

a smoothing procedure of the measuring device we can look at the compensated spectra, $\omega^\beta E(\omega)$ (Fig. 3).

A characteristic height (or length scale), corresponds to the peaks in the compensated spectra. For wind speed spectra averaged over all subsamples the frequencies to the left of the peaks correspond to Kolmogorov and to the right Bolgiano-Obukhov scaling respectively. In addition to high and low frequency scaling anisotropy the UM parameters exhibit component wise anisotropy over high frequencies. The extent of the anisotropy leads to a majority of cases where u_x scales as Kolmogorov and u_y scales as Bolgiano-Obukhov (see estimates at 10m in Table 1). We can further confirm this by comparing the high frequency spectral slopes of the decomposed component vectors (see Fig. 4).

Structure Functions. Using the UM parameters estimated on the wind speed time increments, $\Delta u(\tau) = u(t+\tau) - u(\tau)$, we can test the validity of Taylor's hypothesis by reconstructing a semi-analytic structure function, $\zeta_\tau(q)$, and the empirical structure function, $\zeta_{\mathbf{r}}(q)$, for the temporally averaged wind speed space increments, $\Delta u(\mathbf{r}) = u(\mathbf{x}+\mathbf{r}) - u(\mathbf{x})$:

$$\zeta_\tau(q) = qH - K(q) = qH - \frac{C_1}{\alpha - 1}\left(q^\alpha - q\right) \quad \& \quad \zeta_{\mathbf{r}}(q) = \log\langle|\Delta u(\mathbf{r})|^q\rangle / \log \mathbf{r}, \quad (1)$$

where q is the order of moment and the operator $\langle\cdot\rangle$ denotes an averaging procedure in time. Fig. 5a illustrates that $\mathbf{x} = (x_i)$ is the position vector with reference $i \in [4 : 20]$ (see [3]) and $\mathbf{r}$ is the corresponding spacing between. In Figs. 5b-d, α and C_1 are fixed at 1.5 and 0.35 respectively and only the parameter, H, changes between the two theoretical values 1/3 (red) and 3/5 (blue). The hypothesis is valid when $\zeta_\tau(q) \approx \zeta_{\mathbf{r}}(q)$.

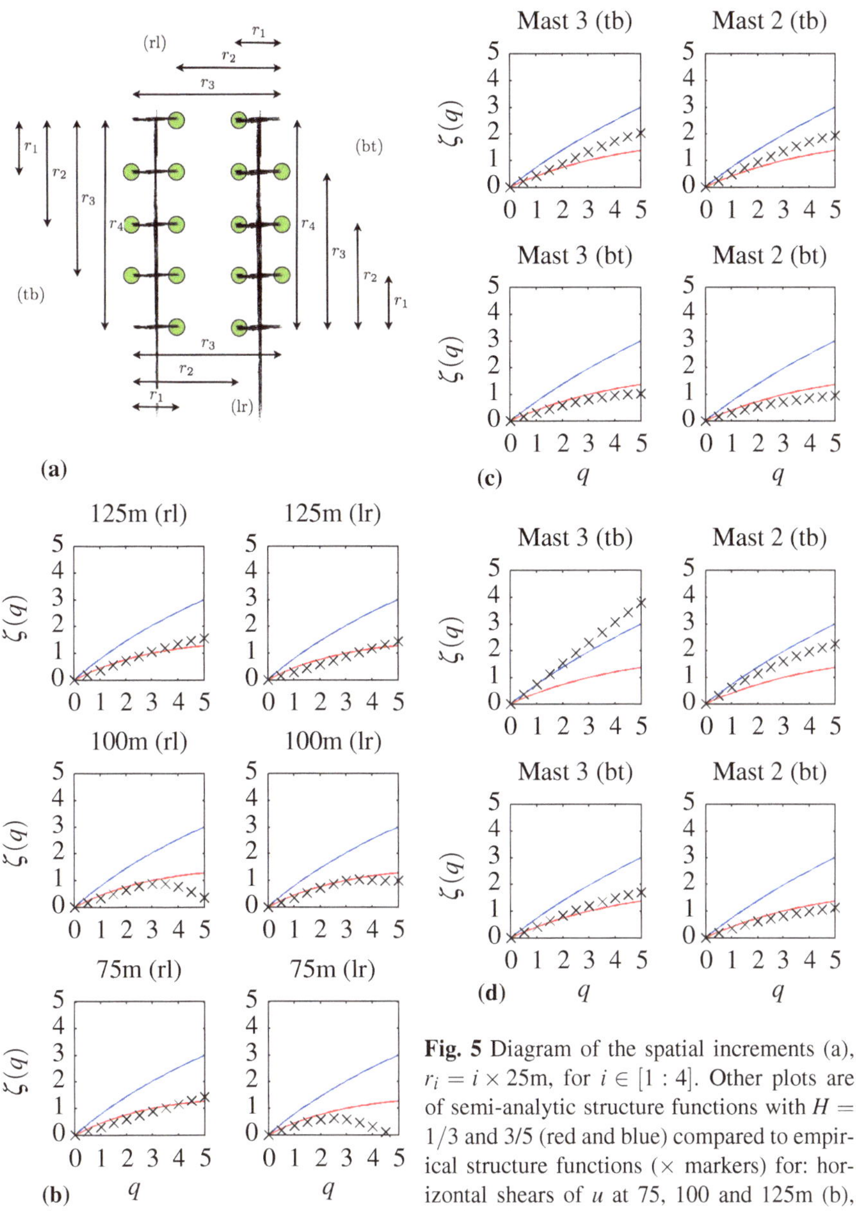

Fig. 5 Diagram of the spatial increments (a), $r_i = i \times 25\text{m}$, for $i \in [1:4]$. Other plots are of semi-analytic structure functions with $H = 1/3$ and 3/5 (red and blue) compared to empirical structure functions ($\times$ markers) for: horizontal shears of u at 75, 100 and 125m (b), vertical shears of u_x-component, (c) and vertical shears of u_y-component (d).

The plots of the structure functions in Fig. 5b confirm Taylor's hypothesis for the horizontal shears of u. Looking at the vertical shears of components, u_x and u_y, we find the anisotropic scaling displayed in the Hurst exponents in Table 1 (estimated on the high frequencies of the time-series velocity increments) and the spectra in Fig. 4 are replicated only for the shears taken from top to bottom (5a). This can be explained with the help of Fig. 3 and the minimum spacing length, $r_1 = 25$m. Again, taking the mean wind speed as 10m/s, we see that taking the smallest vertical shear at the top of the mast, means we are within the Kolmogorov scaling range and taking the smallest vertical shear at the bottom of the mast means we are within the Bolgiano-Obukhov scaling range.

Perspectives for the Modelling of Extremes. The effect of the mean codimension can be seen through the $c(\gamma)$ function (Fig. 6); related to the probability distribution of a flux, ε_λ through:

$$\Pr(\varepsilon_\lambda \geq \lambda^\gamma) \simeq \lambda^{-c(\gamma)}, \quad \text{where} \quad c(\gamma - H) = C_1 \left(\frac{\gamma(\alpha - 1)}{C_1 \alpha} + \frac{1}{\alpha} \right)^{\frac{\alpha}{\alpha - 1}}.$$

Here, $\gamma = \log_\lambda \varepsilon_\lambda$, is a scale invariant singularity. In addition to being able to reproduce the theoretical $c(\gamma)$ curves (Fig. 6) from the UM parameters, the exponent, q_D, of the power law of the distribution of the extremes can be predicted from $K(q_D) = D(q_D - 1)$ with $D = 1$ for a time-series. Over small scales this gives $q_D \approx 4$ and 2.5 for $C_1 = 0.35$ and 0.42 when $\alpha \approx 1.5$, thus defining the extreme behaviour of the energy flux. This further implies $q_D \approx 7.5$ for the wind velocity, which is a crucial result for applications within the field of wind energy.

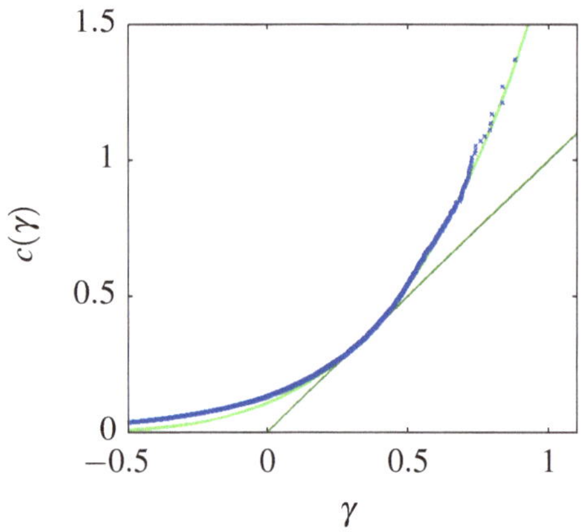

Fig. 6 Theoretical conservative function, $c(\gamma)$, with $\alpha = 1.35$ and $C_1 = 0.35$ for predominantly Kolmogorov scaling (green) compared to the empirical function (blue crosses) using the Weibull plotting position. The empirical data are the concatenated 13 subsamples of the third power of the u_y velocities from Fig. 2b, filtered such that $H < 0.33$. The horizontal shifting of the empirical curves is done so that they superimpose with the theoretical ones at $c(C_1) = C_1$. The shift (corresponding to $H = 0.2$) is comparable to the mean of the first order structure function estimates ($H = 0.25$, see also Fig. 2a).

Acknowledgements. This research is part of the EU-FP7 ITN WAUDIT project.

References

1. Schertzer, D., Lovejoy, S.: Multifractals, Generalized Scale Invariance and Complexity in Geophysics. International Journal of Bifurcation and Chaos 21(12), 3417–3456 (2011)
2. Fitton, G., Tchiguirinskaia, I., Schertzer, D., Lovejoy, S.: Scaling of Turbulence in the Atmospheric Surface-Layer: Which Anisotropy? Journal of Physics: Conference Series 318, 072008 (2011)

3. Palz, W.: European Wind Energy Conference 1984. In: Proceedings of an International Conference Held at Hamburg, FR Germany, October 22-26. Information Today Inc. (1984, 1985)
4. Mücke, T., Kleinhans, D., Peinke, J.: Atmospheric turbulence and its influence on the alternating loads on wind turbines. Wind Energy 14, 301–316 (2011)

Advanced Statistical Analysis of High-Frequency Turbulent Pressure Fluctuations for On- and Off-shore Wind

Andreas Jeromin and Alois Peter Schaffarczyk

Abstract. The turbulence in atmospheric wind was measured on-shore and off-shore with a piezo electric pressure sensor. This sensor operates at a high sampling rate (50 kHz) and survives the corrosive environment in the open sea. Statistical characteristics were investigated for 119 time series and compared to findings in literature. The shape factor s_p^2 for the superposition of distributions was computed and differences to literature were uncovered. With the results for pressure fluctuations we can show that pressure increments are distributed non-Gaussian in gusty wind in some time intervals. So, turbulence characteristics seem to be transient. Furthermore, Taylor micro scale and Kolmogorov scale were estimated for pressure data. The results were reasonable and in range of other experiments. This study is part of ongoing investigations at University of Applied Sciences Kiel.

1 Introduction

For optimization of wind turbine blades focus was laid on laminar to turbulent transition. Among other possibilities transition can be triggered by so called Tollmien-Schlichting waves-packets [12] at high frequencies > 1kHz. Therefore, special emphasis was given to investigations on the turbulent fluctuations in the atmospheric wind.

Andreas Jeromin
Forschungs- und Entwicklungszentrum FH Kiel GmbH,
Schwentinestraße 24, 24149 Kiel, Germany
e-mail: andreas.jeromin@fh-kiel.de
http://www.fh-kiel-gmbh.de

Alois Peter Schaffarczyk
University of Applied Sciences Kiel, Grenzstraße 3, 24149 Kiel, Germany
e-mail: alois.schaffarczyk@fh-kiel.de
http://www.fh-kiel.de

M. Hölling et al. (eds.), *Wind Energy – Impact of Turbulence*,
Research Topics in Wind Energy 2,
DOI: 10.1007/978-3-642-54696-9_9,

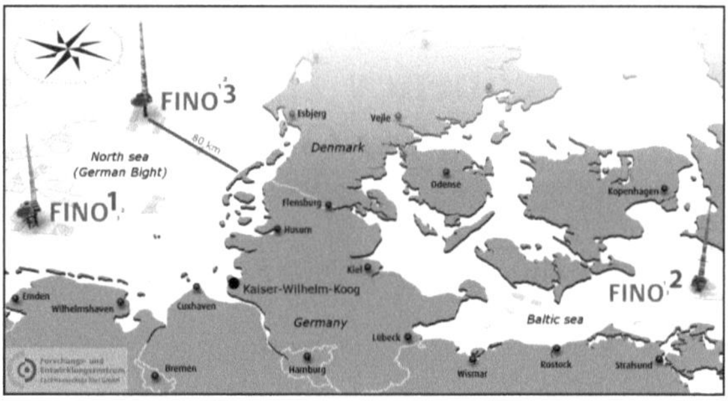

Fig. 1 Locations of measurements on-shore at Kaiser-Wilhelm-Koog test site and off-shore at FINO3 platform 80 km behind the island of Sylt ©FuE-Zentrum FH Kiel GmbH, Graphics: Bastian Barton

For high sampling rates hot wire anemometers are commonly used, but very few measurements were done in the atmosphere [4]. For our measurements a sensor was necessary

- with a high sampling rate (up to 10 kHz) and the capability to resolve the range of interest,
- that could resist the corrosive environment, especially off-shore,
- that could operate autonomous on an unmanned off-shore platform.

It was decided to use a pressure sensor. This sensor was validated against a hot wire anemometer in the wind tunnel at Carl-von-Ossietzky University in Oldenburg, Germany. When properly scaled both sensors showed the same power-spectrum up to 4 kHz [8].

In 2008 on-shore measurements were performed at Kaiser-Wilhelm-Koog test site of Windtest, now DNV-Germanischer Lloyd/Garrad Hassan (GL/GH). The off-shore measurements were carried out also by GL/GH in 2010 on the unmanned scientific platform FINO3 (see fig. 1).

In this study advanced statistical characteristics are analyzed and compared to findings in the literature. It is a sequel of ongoing work at University of Applied Sciences Kiel on turbulent fluctuations in on- and off-shore wind [6, 9, 10, 11].

2 Experimental Setup

For the measurements an ICP pressure sensor (fig. 2, left) from PCB Piezotronics was attached to an imc Meßsysteme GmbH CS-1208 data logger. The sensor had

Fig. 2 Pressure sensor after six month in off-shore service (left) and installation of two parallel pressure sensors T1 and T2 on FINO3 platform (right)

a diameter of 15 mm[1], a pressure resolution of 0.13 Pa and a time resolution of 2.5 Hz up to 80,000 Hz. The pressure data was sampled at 50 kHz. Meteorological data like wind speed and direction were available besides the pressure sensor data and sampled at 1 Hz. Each individual time series was 100 seconds long.

For the on-shore measurements the pressure sensor was placed on a lattice tower 60 m above ground level. The tower stood in Kaiser-Wilhelm-Koog test site of GL/GH (see fig. 1). The time series were categorized by wind directions from land and sea side and by wind speeds of about 6 m/s, 12 m/s and 16 m/s.

The off-shore measurements took place on the unmanned, scientific measurement platform FINO3 and were carried out by GL/GH, too. At a height of 100 m above mid sea level two parallel pressure sensors were mounted as in fig. 2 (right). The considered wind speeds were 6 m/s, 10 m/s, 12 m/s and 16 m/s.

3 Methods Used in Analysis

The data sets were analyzed for power spectral density, probability density distributions and extended self-similarity method of Benzi et al. [2]. In this study the results with the superstatistics from Beck [1] are shown. By superposing a Gaussian distribution with another distribution a shape parameter s^2 can be computed for velocity increments u. This was directly transferred to s_p^2 for pressure increments p as

$$s^2 = \ln\left(\frac{1}{3}\frac{\langle u^4\rangle}{\langle u^2\rangle^2}\right) \quad \Rightarrow \quad s_p^2 = \ln\left(\frac{1}{3}\frac{\langle p^4\rangle}{\langle p^2\rangle^2}\right). \tag{1}$$

From Hölling [5] we saw that s^2 is a shape factor that indicated a Gaussian distribution for $s^2 \approx 0$ and a non-Gaussian distribution otherwise.

A Taylor micro scale λ_p could be estimated from the pressure time series. Comprehensible results were found according to the method derived from Renner [7] with a basic λ_0 computed as

$$\lambda_0^2 = 2\frac{\langle p^2\rangle}{\langle (dp/dt)^2\rangle} \quad ; \quad \frac{dp}{dt} = \frac{p_{i+1} - p_i}{1/\nu_{sampling}} \tag{2}$$

and then the time derivative was replaced by $\frac{dp}{dt} = \lim_{\Delta t \to 0} \frac{p(t+\Delta t)-p(t)}{\Delta t}$. Now λ_p was a function of Δt as shown in fig. 6. A parabola could be fitted to the data points in fig. 6 and the value for λ_p at $\Delta t = 0$ was computed.

The turbulent dissipation rate $\langle \varepsilon_p \rangle = 15\nu \langle u^2 \rangle / \lambda_p$ was estimated by using a Bernoulli ansatz $\langle p^2 \rangle = \left(\frac{\rho}{2}\langle u^2\rangle\right)^2$ to replace $\langle u^2 \rangle$ for pressure. Finally, a Kolmogorov scale $\eta_p = (\nu^3/\langle \varepsilon_p \rangle)^{1/4}$ could be estimated as well, ν being the kinematic viscosity of air ($1.5 \cdot 10^{-5}\ m/s^2$).

[1] A naive estimate of frequency resolution with reference to rated wind-speed may indice an upper limit less than 1 kHz. Because coherence across to wind is in the order of meters we expert all measured fluctuations to be in phase.

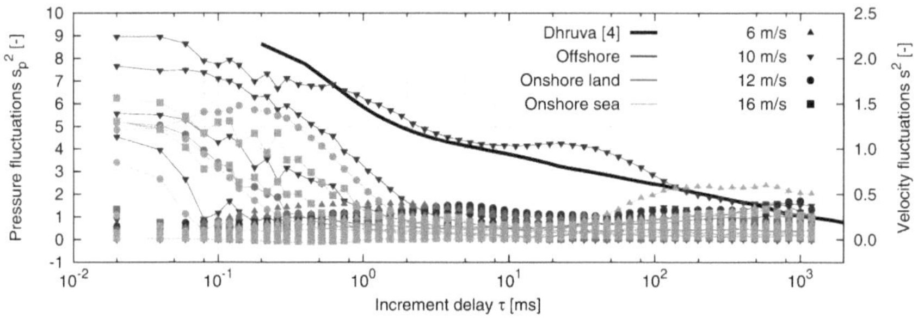

Fig. 3 Shape factor s_p^2 from equation (1) for all measurements on-shore and off-shore. Symbols represent wind speed classes, grayness represents location and wind direction.

4 Results

In previous studies [6, 8] the statistical properties of one on-shore and one off-shore data set were compared at similar wind conditions. Some differences were seen for on- and off-shore wind, but most interesting was that the shape factor s_p^2 was not similar to any other atmospheric experiments [3, 4] or to Continuous-Time-Random-Walk (CTRW) simulations [8].

So all 119 time series were processed and plotted with the shape factor s_p^2 over the increments delay τ in fig. 3. For comparison the shape factor s^2 for velocity measurements from Dhruva [4] with average wind speed of 8.3 m/s is also included. For 110 data sets the shape factor is $s_p^2 \approx 0$ and only nine data sets have a similar behavior compared to Böttcher et al. [3]. A correlation was seen neither to wind speed nor to location.

It was astonishing that four out of the nine data sets with $s_p^2 \neq 0$ were captured within a few minutes in the morning of October 19th, 2010. Nine data sets were available for that day which were joined to form a single data set without gaps at the joints. The wind speed for that day is shown in fig. 4 over measurement time.

The gray boxes in fig. 4 mark 20 s long subsets where the shape factor s_p^2 was evaluated that is shown in fig. 5. At $t = 20$ s the shape factor was $s_p^2 \gg 0$. It dropped to $s_p^2 \approx 0$ only 60 s later at $t = 80$ s. When a gust was passing at $t = 280$ s the shape factor was not affected. A rise in s_p^2 was first observed some time after an other gust passed at $t > 650$ s.

For an other day, August 16th, 2010, ten data sets could be joined, too. The wind speed for that day is also shown in fig. 4. Compared to Oct. 19th the wind was continuous and not gusty. Here, the shape factor was $s_p^2 \approx 0$ throughout.

As explained in section 3 the Taylor micro scale λ_p for Oct. 19th was estimated from the pressure measurements with a parabola fit (fig. 6). For each 100 s data set from Oct. 19th and Aug. 16th the Taylor micro scale λ_p is plotted over the average wind speed $\langle U_{WS} \rangle$ in fig. 7. The variation of λ_p with standard deviation of the wind speed is included as error bars.

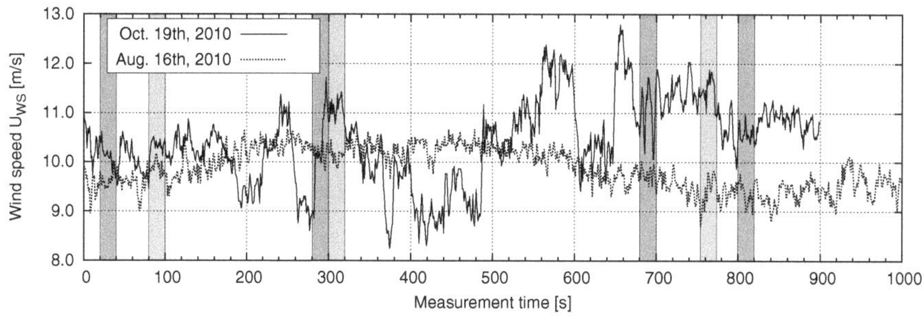

Fig. 4 Wind speeds from joint Offshore measurements on Aug. 16th and Oct. 19th, 2010

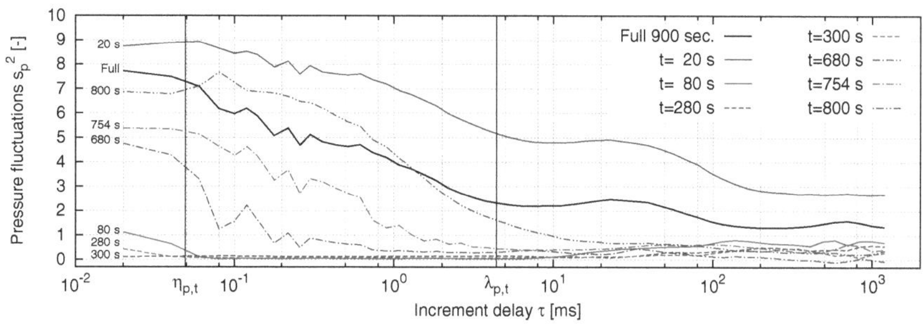

Fig. 5 Shape factors s_p^2 on Oct. 19th, 2010 at specified subsets from fig. 4

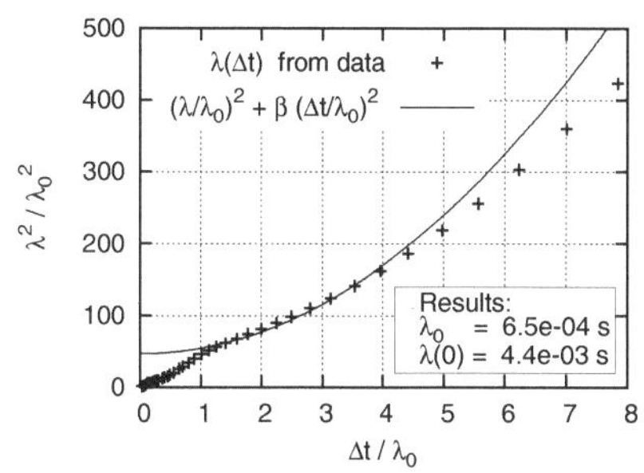

Fig. 6 Estimating Taylor micro scale λ_p for joint data set from Oct. 19th, 2010. Index p,t for λ was omitted.

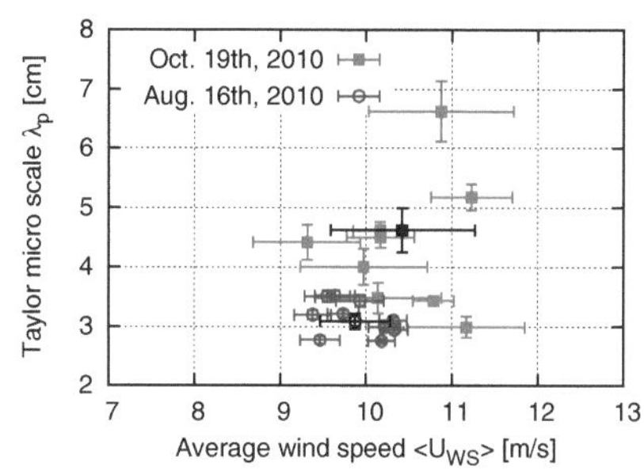

Fig. 7 Taylor micro scale for 100 s measurements estimated from pressure data

The values for Aug. 16th focus in a close area in fig. 7. On Oct. 19th the wind was gusty and strong variations in wind speed appeared. This reflects in Taylor micro scale λ_p that is scattered over a wide range. The error bars are larger, too.

For the joint data set from Oct. 19th the Taylor micro scale λ_p was also computed. The fitting in fig. 6 returned $\lambda_{p,t} = 4.4$ ms. With an average wind speed $\langle U_{WS} \rangle = 10.4$ m/s, the Taylor length was estimated as $\lambda_p = \langle U_{WS} \rangle \, \lambda_{p,t} = 4.62$ cm. The value for joint data from Aug. 16th was $\lambda_p = 3.09$ cm. Both values are included in fig. 7 in black. The estimated values for λ_p were in the range of other experiments [3].

The Kolmogorov scale η_p was estimated for the joint data sets on Oct. 19th to be $\eta_p = 0.51$ mm, for Aug. 16th it was $\eta_p = 0.23$ mm. These values were in range of other experiments [3], too.

$\eta_{p,t}$ and $\lambda_{p,t}$ are also marked in the shape factor graph (fig. 5) for Oct. 19th.

5 Conclusions and Outlook

With our results of analyzing pressure fluctuations we can show that the distribution of pressure increments are Gaussian for most data sets. Only nine out of 119 measurements have a pronounced non-Gaussian distribution. The distributions of pressure increments were Gaussian for steady wind, but not for gusty wind in some time intervals. So, turbulence characteristics seem to be transient.

Furthermore, Taylor micro scale and Kolmogorov scale were estimated for pressure data. The results were reasonable and in range of other experiments.

Next, other combined data sets will be evaluated. There is a chance to find a possible correlation to gustiness of the wind for the non-Gaussian distributions or to boundary layer stability. Also, experiments are to be carried out with a new Laser-Cantilever-Anemometer for atmospheric applications, that is currently developed at Carl-von-Ossietzky University Oldenburg.

Acknowledgements. Our thanks go to Detlef Stein and Bastian Schmidt from Germanischer Lloyd/Garrad Hassan for setting up the probes on-shore and off-shore and performing the data acquisition. We also thank Dr. Michael Hölling from Carl-von-Ossietzky University Oldenburg for wind tunnel testing of the sensors. The project was funded by German Federal State Schleswig-Holstein. (Funding Contract No. 122-09-023).

References

1. Beck, C.: Superstatistics in hydrodynamic turbulence. Phys. D 193, 195–207 (2004)
2. Benzi, R., Ciliberto, S., Tripiccione, R., Baudet, C., Massaioli, F., Succi, S.: Extended self-similarity in turbulent flows. Phys. Rev. E 48, R29–R32 (1993)
3. Böttcher, F., Barth, S., Peinke, J.: Small and large scale fluctuations in atmospheric wind speeds. Stoch. Environ. Res. and Risk Assess. 21, 299–308 (2007)
4. Dhruva, B.R.: An Experimental Study of High Reynolds Number Turbulence in the Atmosphere. Yale Univ. Conn., USA (2000)
5. Hölling, M.: Sensorentwicklung für Turbulenzmessungen. Carl-von-Ossietzky Univ. Oldenbg., Ger. (2008)
6. Jeromin, A., Schaffarczyk, A.P.: Measurement of high frequency pressure fluctuations at german offshore platform FINO3. In: EWEA Offshore 2011 Conf. & Exhib., Amst., The Neth. (2011)
7. Renner, C.: Markowanalysen stochastisch fluktuierender Zeitserien. Carl-von-Ossietzky Univ. PHD thesis, Oldenburg, Gerermny (2002)
8. Schaffarczyk, A.P., Jeromin, A.: unpublished internal Report No. 78, Kiel, Germany (2011)

9. Schaffarczyk, A.P., Schwab, D.: Offshore aerodynamic turbulence and its relevance to blade design. In: Eur. Offshore Wind Conf. & Exhib., Berl., Ger. (2007)
10. Schwab, D., Schaffarczyk, A.P.: Comparing on- and off-shore aerodynamic turbulence by measurement. In: Eur. Offshore Wind Conf. & Exhib., Stockh., Swed. (2009)
11. Schwab, D., Schaffarczyk, A.P.: Advanced investigation into the influence of high frequency turbulence on wind turbine. In: Voutsinas, S., Chaviaropoulos, T. (eds.) The Sci. of Mak. Torque from Wind 2010, Heraklion, Crete, Greece, pp. 219–225 (2010)
12. Reeh, A.D., Weissmüller, M., Tropea, C.: Free-Flight Investigation of Transition under Turbulent Conditions on a Laminar Wing Glove. In: Proc. 51st AIAA Sciences Meeting, Grapevine, TX, USA (2013)

Characterizing the Fluctuations of Wind Power Production by Multi-time Statistics

Oliver Kamps

Abstract. The fluctuations of electrical energy, generated by wind turbines, reflect the interaction between the turbulent wind field and a complex technical system. In this article we study time series of the integrated wind power production of a large number of spatially distributed wind energy converters. The aim is to characterize the fluctuations of wind power production on different time scales.

1 Introduction

Wind energy is a steadily growing and highly fluctuating source of energy [1]. The wind energy production of a whole country is a result of the complex interaction between atmospheric turbulence and a large number of wind power generators at different places. The complexity of the interaction between weather dynamics and the wind farms is compressed into the summarized energy production from all wind farms. In this article we want to make a first step to explore the statistics of this observable on multiple time scales.

From recent research we know that the fluctuating character of the atmospheric wind is transferred into the power production of single wind turbines and wind parks [2]. On this level the energy production is characterized by extreme fluctuations. To contrast these results with the production of a whole country, we analyze a data set of the summarized wind power generation of Ireland [3]. The focus lies on the increments of the power generation since they carry the information about the fluctuations. The fluctuations are the hallmark of wind energy and distinguish this energy resource from conventional energy resources like coal or atomic power which are mainly used to produce energy at a constant rate.

Oliver Kamps
Center for Nonlinear Science, Universität Münster, Correnstraße 2, Münster, Germany
e-mail: okamp@uni-muenster.de

M. Hölling et al. (eds.), *Wind Energy – Impact of Turbulence*,
Research Topics in Wind Energy 2,
DOI: 10.1007/978-3-642-54696-9_10,

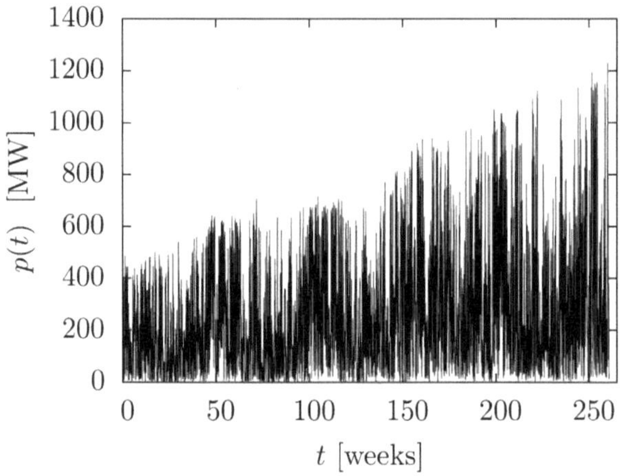

Fig. 1 Data series of the wind power generation in Ireland from 01.01.2006 to 31.12.2010 [3]

2 The Data Set

The data set consists of $\approx 2 \cdot 10^5$ measurements of the whole wind energy production of Ireland between 01.01.2006 and 01.01.2011. Every data point is computed by averaging the power generation over 15 minutes, which leads to a sampling interval of Δt =15min for the signal. To get a first impression of the signal in Fig. 1, the whole data set is shown. Beside the strong fluctuations one immediately observes the systematic trend in the data, which shows that over the years more and more wind power generation capacity has been installed in Ireland.

While a systematic trend of the mean of the signal does not affect the increment statistics, the growth of the standard deviation does. Since we want to use the whole data set to investigate the increment statistics, we have to apply a detrending procedure to disentangle the fluctuations and the trend of the standard deviation. The hypothesis underlying this approach is that we assume that the fluctuations obey the same form of probability density function (PDF) everywhere in the signal. This assumption was checked by estimating the PDF of the increments $p(t+\tau) - p(t)$ for $\tau = \Delta t$ over time windows of 16 weeks length and comparing the rescaled PDFs to each other.

Motivated by this we divide the data set into intervals of 16 weeks and compute for every interval the standard deviation $\sigma(t)$ of the signal. Plotting this data against time, we observe a linear trend in the evolution of the standard deviation. This trend can be fitted very well by the linear function $\sigma(t) = at + b$ with $a = 0.00087$ and $b = 117$. The detrended data set $x(t)$ is then computed by

$$x(t) = \frac{p(t)}{at+b} \,. \tag{1}$$

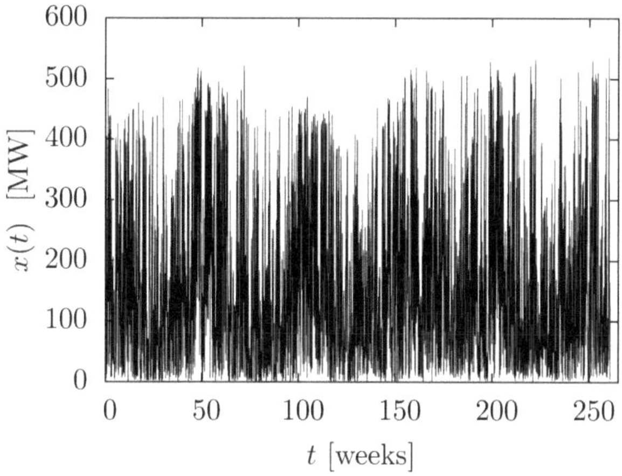

Fig. 2 Same data series as in Fig. 1 but rescaled by the procedure described in the text

The result is depicted in Fig. 2. To check the sensitivity of this detrending procedure on the choice of the time window, it was repeated for time intervals ranging from 8 weeks to 24 weeks, leading to the result that a and b do not vary significantly with respect to the chosen interval.

3 Characterizing the Fluctuations

In order to analyze the fluctuations, we now consider the increments

$$s(\tau) = x(t+\tau) - x(t) \tag{2}$$

of the detrended power generation time series $x(t)$. In Fig. 3 the rescaled ($\sigma = 1$) PDF of the increments for $\tau = \Delta t$ is shown (the rescaled variable is denoted by v) together with a Gaussian PDF and a Laplace distribution

$$f(v) = \frac{1}{\sqrt{2}} e^{-\sqrt{2}|v|}. \tag{3}$$

Except from the far tails, where the estimation of the PDF is not reliable, the increment PDF can be extremely well described by the Laplace distribution. This immediately has important consequences for the occurrence of very large fluctuations. Compared to a Gaussian PDF a 5σ event (corresponding to a 95 MW fluctuation in the detrended data set) is about 400 times more likely to occur. This shows that even on the scale of a large number of wind turbines that are distributed over Ireland, the increment PDF is far from being Gaussian and has big tails. For the practical integration of wind energy, this means that large power fluctuations on very short time scales have to be managed.

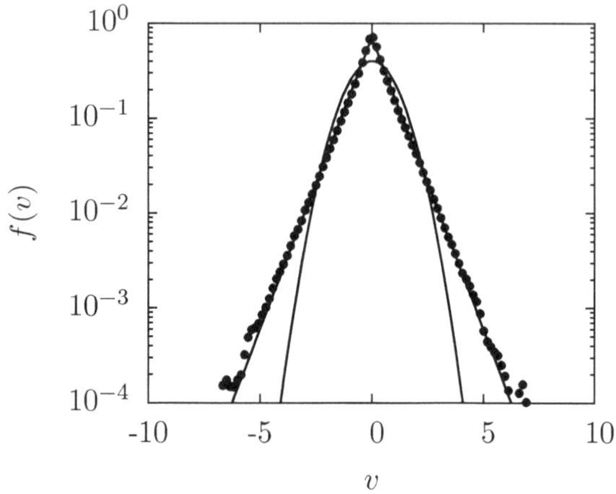

Fig. 3 Semi-logarithmic plot of the rescaled increment PDF for $\tau = \Delta t$ (15min) together with a Lorentz (line) and a Gaussian (parabola) distribution

In Fig. 4 several non-rescaled increment PDFs up to a time lag $\tau = 672\Delta t$ corresponding to 1 week are shown. One can see that on larger time scales the tails are less pronounced but the PDFs are still far from Gaussian, especially in the central part. Even on large time scales the fluctuations are extreme compared to Gaussian fluctuations.

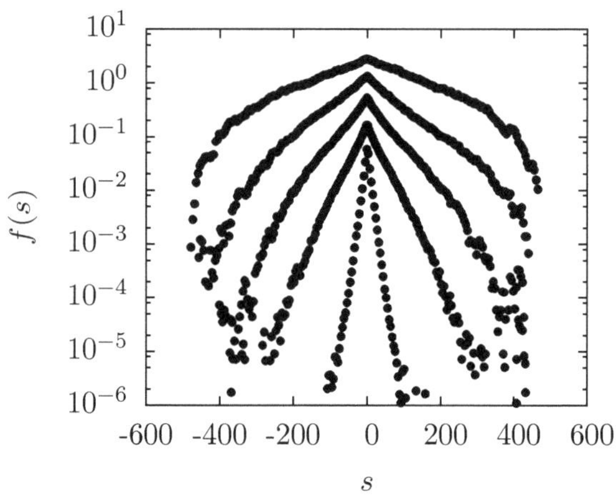

Fig. 4 Semi-logarithmic plot of the non-rescaled increment PDFs for $\tau = \Delta t$ (15min), $12\Delta t$ (3h), $24\Delta t$ (6h), $48\Delta t$ (12h), $672\Delta t$ (1week) (from inner to outer PDF)

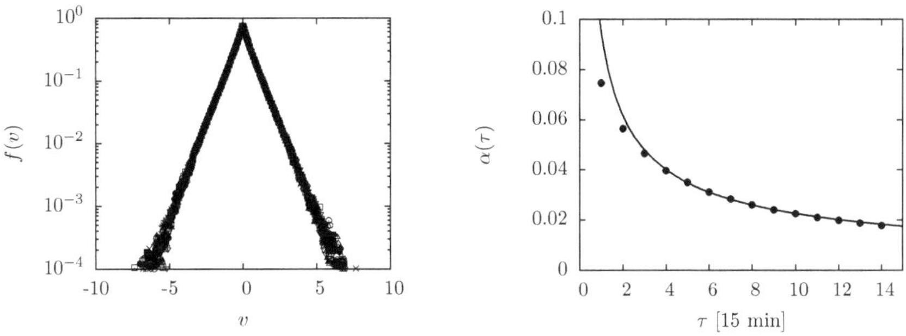

Fig. 5 Left: Semi-logarithmic plot of the rescaled increment PDFs for $\tau = \Delta t$ (15min), $2\Delta t$ (0.5h), $\ldots$, $10\Delta t$ (2.5h). Right: Fit parameter $\alpha(\tau)$ (dots) together with $a\tau^{-b}$ (line).

Plotting the rescaled increment PDFs for τ between 15 min and 2.5 h shows that the increment PDFs in this range are self-similar since all PDFs collapse on one curve (see left part of Fig. 5). The non-rescaled PDFs can be fitted by

$$f(v,\alpha) = \frac{\alpha}{2} e^{-\alpha|v|} \tag{4}$$

with a time-lag dependent coefficient $\alpha(\tau)$. This coefficient is shown in the right Part of Fig. 5 together with the power law $a\tau^{-b}$ with $a = 0.095$ and $b = 0.625$ as a guide for the eye. Although the first two values deviate from this power law, it seems to be a very good parametrization for the largest part of the self-similar range. For larger time lags than 2.5h, the increment PDFs loose self-similarity and cannot be parametrized in this simple way.

4 Conclusion and Outlook

The central result of the present paper is the fact that even in the case of a large number of wind power generating units spread over a country, the summarized statistics of the increments is still far from Gaussian. Even for time lags in the order of days this non-Gaussianity remains. A possible explanation of this observation is that the integral length of the atmospheric turbulence is in order of several hundreds of kilometers [5]. For this reason the output of different wind turbines at different places in the country is highly correlated which significantly slows down the convergence of the PDF for the summarized signal to a Gaussian. Even for countries much larger than Ireland this should play an important role.

Since it does not seem possible to circumvent extreme fluctuations in wind power generation, it is necessary to understand and model these fluctuations. Therefore the next step will be to construct a model that captures the main features of the stochastic evolution of the wind energy production. The modeling process could be assisted by the methods presented in [4]. Such a model can be used as a base for the short term prediction of wind energy production or as input for research on power grid design and stability in the context of the grid integration of wind energy.

References

1. MacKay, D.JC.: Sustainable Energy - Without the Hot Air, 1st edn. UIT Cambridge Ltd. (February 2009)
2. Wächter, M., Heisselmann, H., Hölling, M., Morales, A., Milan, P., Mücke, T., Peinke, J., Reinke, N., Rinn, P.: The turbulent nature of the atmospheric boundary layer and its impact on the wind energy conversion process. Journal of Turbulence, N26 (2012)
3. Data freely, `http://www.iregrid.com`
4. Friedrich, R., Peinke, J., Sahimi, M., Tabar, M.R.R.: Approaching complexity by stochastic methods: From biological systems to turbulence. Physics Reports 506(5), 87–162 (2011)
5. Baile, R., Muzy, J.-F.: Spatial intermittency of surface layer wind fluctuations at mesoscale range. Phys. Rev. Lett. 105, 254501 (2010)

Wind Energy: A Turbulent, Intermittent Resource

Patrick Milan, Allan Morales, Matthias Wächter, and Joachim Peinke

Abstract. Statistics of high-frequency wind speed and power output measurements display turbulent fluctuations. Frequent wind gusts are observed through heavy-tailed statistics of increments of the wind velocity. These complex statistics cannot be reproduced using Gaussian wind field models, stressing the need for appropriate turbulence models. Wind fluctuations and wind gusts are converted by a wind energy converter (WEC) into electrical power fluctuations and power gusts fed into the grid. Similar heavy-tailed statistics are observed for the power output of a single WEC, a wind farm and to some extend to a large array of WECs. Furthermore, a stochastic approach is presented to model the conversion process operated by a WEC. Our results illustrate the impact of turbulence on wind power production, stressing the importance of turbulence research to properly integrate more and more wind energy into electrical networks.

1 Introduction

The ongoing development of wind energy presupposes major changes in energy strategy in the coming decades. While the installed wind power capacity grows rapidly, rise the questions of wind power prediction and grid stability. While such questions are commonly addressed for slow, meteorological scales, fast turbulent dynamics are still an open topic of research. The impact of turbulence on the wind energy conversion process is essentially a high-frequency problem, that must be addressed as such. We focus in section 2 on the turbulent statistics of atmospheric wind measurements at time scales of seconds to minutes. This analysis is extended in section 3 to the power output of a wind energy converter (WEC), than can be remodeled

Patrick Milan · Allan Morales · Matthias Wächter · Joachim Peinke
ForWind - Center for Wind Energy Research, Oldenburg
e-mail: {patrick.milan,allan.morales,matthias.waechter,
joachim.peinke}@forwind.de

M. Hölling et al. (eds.), *Wind Energy – Impact of Turbulence*,
Research Topics in Wind Energy 2,
DOI: 10.1007/978-3-642-54696-9_11,

using a stochastic approach. The impact of turbulent fluctuations on grid stability issues is then addressed in section 4 where the statistical analysis is presented for a wind farm and for a large array of WECs in Ireland. All results presented within this paper were extracted from data presented in table 1.

Table 1 Datasets used with corresponding information and naming convention

Dataset	Measurement duration	Sampling rate	Wind speed	Power output
FINO	6 months	1 Hz	u	-
WEC	2 weeks	1 Hz	u_{WEC}	P_{WEC}
wind farm	3 weeks	1 Hz	u_{FARM}	P_{FARM}
Eirgrid (2005 to 2009)	4 years	1/900 Hz	-	P_{GRID}

When addressing the impact of turbulence on wind energy converters (WECs), one should stress the influence and necessity of international norms and standards. These standards offer a common ground from which wind energy applications can be performed, ranging from the assessment of wind potential [2] to WEC performance [3], fatigue estimations and so forth. The use of ten-minute averaging on measurement data is a common starting point for such applications. While this is a meaningful approach for long-term analysis, fast wind dynamics in the scales of seconds should be accounted as well. They influence fatigue loads sustained by WECs, see [6] and power fluctuations in the grid, as we address here.

2 Statistics of Wind Turbulence

As our work focuses on high-frequency dynamics, time series of wind speed u_{WEC} and power output P_{WEC} are presented in figure 1. The ten-minute averages observed here are only a slowly-changing, coarse estimate of the faster wind and power dynamics. Fast but nonetheless large fluctuations are overseen. Such fluctuations have an obvious impact on the quality of the electrical power fed into the grid, as presented in section 4.

Fast fluctuations are commonly represented by their standard deviation, quantifying their amplitude. But this statistical description (ten-minute average and standard deviation) does not inform on the existence of gusts. However, *gustiness* is a characteristic feature of turbulent flows. A striking example is observed in figure 1 at time 32 min, where a wind gust creates a power gust, the output power raising from 10% to 100% within seconds. Wind gusts are not estimated from the measured wind speed u, but rather observed through the velocity increment $u_\tau(t) = u(t+\tau) - u(t)$ that measures the change in wind speed u over a time scale τ.

Gustiness is related to the probability of observing large increments u_τ, thus indicating a gust. The probability density function (PDF) of some wind increments $p(u_\tau)$ is presented in figure 2(a) for u. The *heavy-tailed* form of the increment PDF

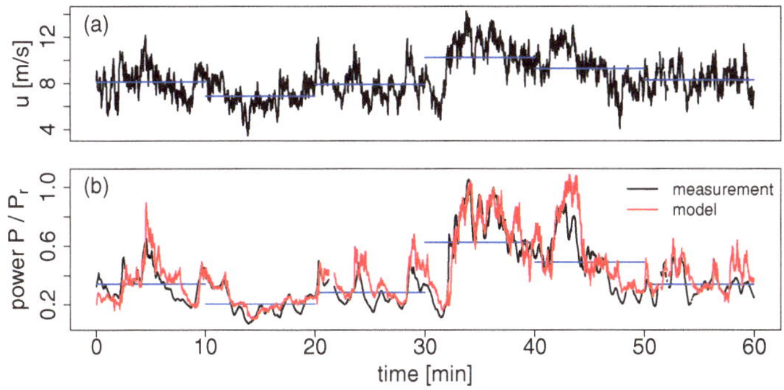

Fig. 1 (a) Excerpt of wind speed u_{WEC} sampled at 1 Hz; (b) simultaneous measurement of the power output P_{WEC} at 1 Hz (black) and corresponding signal (red) modeled by equation (1) as described in section 3.1. Ten-minute averages are indicated for u_{WEC} and P_{WEC} (horizontal blue lines). Power values are normalized by the rated power P_r of the WEC.

indicates that extreme events, i.e. wind gusts are more frequent than what a Gaussian distribution would predict. This is particularly true over short time scales of few seconds, where events up to 20 standard deviations ($u_\tau \simeq 20\sigma_\tau$) are recorded. Concretely, a wind speed change of 3 m/s (10σ) over 1 s occurs every 4 hours for the atmospheric wind measured, against every 5 years for a Gaussian wind field. Ref [5] proposes a detailed statistical analysis of atmospheric turbulence. The impact of such wind gusts on the electrical output is presented in section 3 for a WEC and in section 4 for larger arrays of WECs.

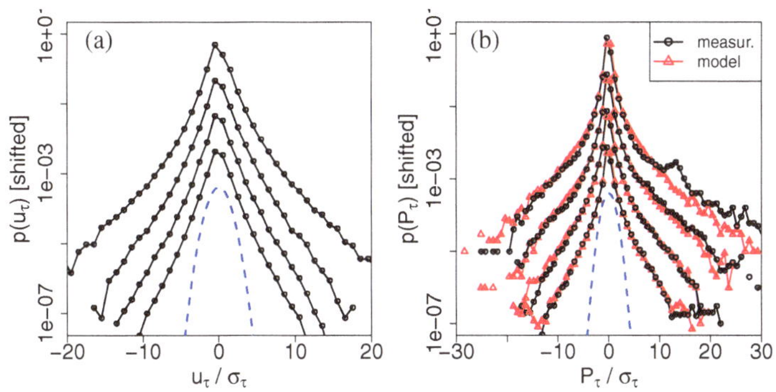

Fig. 2 Increment PDF for time scales $\tau = \{1, 8, 32, 128\}$ s (top to bottom) for (a) velocity increments u_τ calculated from u on the offshore platform FINO; (b) power increments P_τ calculated from P_{WEC} (black circles) and corresponding signal (red triangles) modeled from equation (1). The increments are normalized by their standard deviation σ_τ for comparison with a Gaussian distribution (blue dashed curve). All PDFs are shifted vertically for clarity of reading.

3 Wind Power Output

We see in figure 1 that wind fluctuations are converted by the WEC into power fluctuations. Following our analysis of wind speed increments in section 2, we define the power output increment $P_\tau(t) = P(t+\tau) - P(t)$ as the change in the electrical power output P over a time scale τ. PDFs of power increments are presented in figure 2(b). They appear more heavy-tailed (more intermittent) than for the wind speed. This could be anticipated as the power scales nonlinearly (roughly cubically) with the wind speed. These *power gusts* are relatively minor at the time scale $\tau = 1$ s, as the most extreme events $30\sigma_\tau \simeq 5\%$ rated power. The WEC cannot react significantly over such a short time of 1 s. However, power gusts of $\simeq 40\%$ rated power are observed over $\tau = 8$ s. We observe that less than 30 s is necessary for the power output to change from 0% to 100%, or vice versa. Ref [6] also showed the direct impact of turbulence on fatigue loads, stressing the replacement of Gaussian models by turbulence models.

3.1 Stochastic Model of Power Output

The conversion process $u \to P$ operated by a WEC is commonly described by a power curve [3]. While the IEC power curve is a good model at low frequencies (for example to estimate the annual energy production), it does not suffice to model the high-frequency WEC dynamics as observed in figure 1. A stochastic description is proposed in [1] where the new concept of a Langevin power curve describes the WEC dynamics. Unlike the IEC power curve where ten-minute data is used, the Langevin power curve describes the conversion process at the reaction time of the WEC, typically some seconds. We extend the concept to a stochastic model that reproduces the conversion process $u \to P$ at a high frequency of 1 Hz, see [4]. Knowing the wind speed $u(t)$, the power output $P(t)$ of the WEC is modeled by a stochastic Langevin equation

$$\frac{dP(t)}{dt} = D^{(1)}(P;u) + \sqrt{D^{(2)}(P;u)} \cdot \Gamma(t), \tag{1}$$

where the Kramers-Moyal coefficients $D^{(n)}$ are estimated from two signals $u(t)$ and $P(t)$ measured beforehand

$$D^{(n)}(P;u) = \lim_{\tau \to 0} \frac{1}{n!\tau} \langle [P(t+\tau) - P(t)]^n \mid P(t) = P; u(t) = u \rangle, \tag{2}$$

and $\Gamma(t)$ is a gaussian white noise with mean value $\langle \Gamma(t) \rangle = 0$ and variance $\langle \Gamma(t)^2 \rangle = 2$, see [7] for a theoretical description. The conversion process is separated into a deterministic drift $D^{(1)}$ that represents the WEC dynamics, plus some random fluctuations $\sqrt{D^{(2)}}\Gamma(t)$. A typical time series is shown in figure 1(b), where measured and modeled power output signals are compared. Also, increment statistics can be compared in 2(b). The Langevin model yields intermittent power increments, similarly to what was measured. As such, we promote it as as realistic approach to model WEC dynamics including high-frequency fluctuations and gusts.

4 Grid Integration Challenge

Results presented in section 3 indicate that the power output of a single WEC is intermittent. Wind gusts are converted into power gusts within seconds. This claim is often criticized by the intuition that such power fluctuations are only meaningful for a single WEC. The act of clustering WECs into wind farms or even larger arrays should effectively randomize the fluctuations, such that the total power output is no longer intermittent. We address this claim by computing power increments P_τ for the total power output of a wind farm P_{FARM}, and for the wind power P_{GRID} generated by Eirgrid in Ireland, representing an installed capacity in the order of 1000 MW (data freely available on www.eirgrid.com). Increment PDFs are presented in figure 3. We

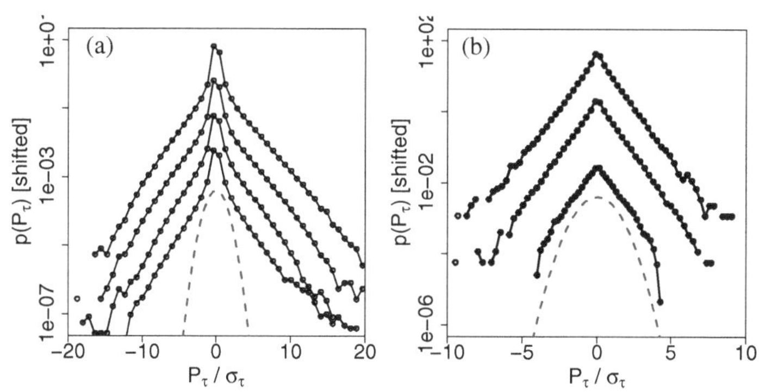

Fig. 3 increment PDFs $p(P_\tau)$ of (a) wind farm power P_{FARM} for time scales $\tau = \{1, 8, 32, 128\}$ s (top to bottom); (b) of total wind power for Eirgrid P_{GRID} for time scales $\tau = \{15$ min,1 hour,2 days$\}$. The increments are normalized by their standard deviation σ_τ for comparison with a Gaussian distribution (blue dashed curve). All PDFs are shifted vertically for clarity of reading.

observe heavy-tailed increment PDFs for the wind farm power, with extreme events of about $20\sigma_\tau$ up to 32 s. These events represent a change of power of about 40% of the farm capacity. For information, power changes up to 90% are recorded within 5 minutes. Intermittency is as well present in the wind power produced in Ireland. The most extreme events recorded over 15 min are about $10\sigma_\tau$, representing a change in power of about 170 MW.

5 Conclusion

Atmospheric wind measured at a high frequency of 1 Hz display complex turbulent fluctuations. Such effects cannot be described using the ten-minute statistics prescribed by the IEC norms, including Gaussian models that underestimate the occurrence of wind gusts. Wind gusts have a central impact on WECs, that convert them into power gusts. This is observed on high-frequency power signals measured. We observe changes in power output of about 50% within 10 s and 100% within 30 s for a single WEC. The wind farm can see its power output change by 50% in about 30 s and by 90% within 5 minutes. The wind farm configuration does filter the fluctuations to some extent, but large wind gusts are nonetheless converted into farm power gusts. Such gustiness is also present on the wind power generated in Ireland. While less intermittent, these gusts represent much larger power fluctuations. We observe the tendency that clustering more WECs together does filter some of the intermittency, but the power fluctuations are much larger in absolute terms.

Intermittency is a central aspect in turbulence research, but it is still seldom addressed in the wind energy community. Large power fluctuations are fed into the electric grid, raising the question of the grid stability in future, largely wind-powered grids. Alternatives should be implemented, possibly in the form of smart controllers to regulate turbulent fluctuations in the grid. We believe that such alternatives are essential to the ongoing integration of wind energy.

Acknowledgements. We thank the German Ministry for Environment and the German Ministry for Education and Research for funding this project. Special thanks go to Edgar Anahua, Stephan Barth, Julia Gottschall, Michael Hölling and Tanja Mücke for their contribution to this work.

References

1. Gottschall, J., Peinke, J.: How to improve the estimation of power curves for wind turbines. Environmental Research Letters 3(1), 015005, 7 (2008)
2. IEC 61400-1, Wind turbines. Part 1: Design requirements. International Electrotechnical Commission, IEC/61400-1, 3rd edn. (2005)
3. IEC 61400-12-1, Wind Turbine Generator Systems, Part 12: Wind turbine power performance testing. International Electrotechnical Commission, IEC/61400-12-1, 1st edn. (2005)
4. Milan, P., Wächter, M., Peinke, J.: Stochastic modeling of wind power production. In: Proceedings of EWEA 2011, Brussels (2011)
5. Morales, A., Wächter, M., Peinke, J.: Characterization of wind turbulence by higher-order statistics. In: Wind Energy (2011) (accepted)
6. Mücke, T., Kleinhans, D., Peinke, J.: Atmospheric turbulence and its influence on the alternating loads on wind turbines. Wind Energy 14(2), 301–316 (2011)
7. Risken, H.: The Fokker-Planck equation. Springer, Berlin (1984)

Highly Time-Resolved Production Data Analyzed Using Engineering Wind-Farm Models

U.V. Poulsen, J. Scholz, E. Hedevang, J. Cleve, and M. Greiner

1 Introduction

Wind farm design and operation require a good understanding of the complex intra-farm wind flows, with the level of sophistication depending on the concrete objectives. Often speed and thorough validation are essential and one prefers to focus on the average flow field using analytical, engineering wake models [6, 5] or highly simplified computational fluid dynamics calculations [1]. However, when standard engineering wind-farm wake models are employed for e.g. the maximization of wind-farm power through collaborative control schemes [8, 3], the results from experiments do not fully match the theoretical prediction. In fact, intra-farm wind flows are very dynamic with wake meandering, wake interaction and the entrainment of kinetic energy from the atmospheric layer above the wind-farm boundary layer. It is desirable take these dynamical effects into account, while still working with descriptions simple enough to be used efficiently in an optimization calculation. As a step in this direction, we here present a methodolgy where high-frequency intra-farm power generation data from the offshore wind farm Nysted is analyzed by comparison to several variations of engineering wake models, which differ in the description of single wakes as well as multiple wake interactions.

U.V. Poulsen · J. Scholz · M. Greiner
Dep. of Engineering, Aarhus University, DK-8000 Aarhus C, Denmark
e-mail: uvp@phys.au.dk, {scholz,greiner}@imf.au.dk

E. Hedevang
Dep. of Mathematics, Aarhus University, DK-8000 Aarhus C, Denmark and
Siemens Wind Power, Borupvej 16, DK-7330 Brande, Denmark
e-mail: emil@imf.au.dk

J. Cleve
Siemens Wind Power, Borupvej 16, DK-7330 Brande, Denmark
e-mail: jochen.cleve@siemens.com

M. Hölling et al. (eds.), *Wind Energy – Impact of Turbulence*,
Research Topics in Wind Energy 2,
DOI: 10.1007/978-3-642-54696-9_12, © Springer-Verlag Berlin Heidelberg 2014

2 Method

The used data set consists of three months (march – may 2009) of recordings from the 72 Siemens-Bonus 2.3 MW turbines comprising the Nysted off-shore wind farm. The temporal resolution of the recordings, including turbine power and yaw, is 1 second. Data from nearby meteorology masts is not used. The 3 month time span of the data is divided into disjoint 10 minutes periods. Only data where the frontline turbines produce more than 150 kW and less than 2150 kW are considered. This filter excludes about 20% of the data because of low wind and another 20% because of high wind. The filter is applied because we want to be able to translate power to wind speed via the inverse power curve [2], which can only be done comfortably between cut-in and rated power. From the filtered 10 minutes periods the first 15, 30, 60, 120, 300, and 600 seconds averaged data points are selected. These form the ensemble of events to be fitted to the models.

We use two models for the far wake expansion. In the Jensen model [6] the wake radius at a distance x behind the turbine is given by a linear expression

$$R_{\mathrm{J}}(x) = R + kx \tag{1}$$

where R is the rotor radius and k is a parameter that determines how fast the wake expands. The effective wind speed felt by a downwind turbine is

$$u_{\mathrm{J}}(x) = v_0\left[1-\left(1-\sqrt{1-C_T(v_0)}\right)\left(\frac{R}{R_{\mathrm{J}}(x)}\right)^2\frac{A_{\mathrm{overlap}}}{A_{\mathrm{rotor}}}\right], \tag{2}$$

where C_T is the thrust coefficient, A_{rotor} is the rotor area πR^2, and A_{overlap} is the area of the overlap between the wake and affected turbine rotor. In what we will refer to as the Frandsen model [5] we instead have square root dependence

$$R_{\mathrm{F}}(x) = R\sqrt{\max\left\{\beta, \frac{\alpha x}{2R}\right\}} \quad \text{for} \quad \beta = \frac{1+\sqrt{1-C_T}}{2\sqrt{1-C_T}}, \tag{3}$$

with the adjustable expansion parameter α and a wake velocity

$$u_{\mathrm{F}}(x) = \frac{v_0}{2}\left[1+\sqrt{1-2C_T(v_0)\left[\frac{R}{R_{\mathrm{F}}(x)}\right]^2\frac{A_{\mathrm{overlap}}}{A_{\mathrm{rotor}}}}\right]. \tag{4}$$

For the Jensen model we follow the Katic approach and treat overlapping wakes by adding squared velocity deficits [7]. For the Frandsen model we use both the Katic approach and a semi-linear approach described in [10]. In addition, for wake affected turbines, the thrust coefficient is evaluated at the estimated facing wind, i.e. v_0 in $C_T(v_0)$ in Eqs. (2) and (4) is replaced by what the wakes from upwind turbines result in.

For each event an approximate wind direction is first determined by a simple average of the yaw readings of all turbines with a correction as described in Ref. [4]. This fixes the most upwind turbines on the farm perimeter. Then an either homogeneous or heterogeneous estimate of the free wind v_0 is found from the inverse-power-curve estimates from these most upwind turbines: In the homogeneous model, there is a single free wind direction and speed for the whole farm, and a simple average is applied. In the heterogenous model, we still assume a single wind direction, but the free wind speed is allowed to vary in the cross-wind direction as a second order polynomial. This ensures that we can include major cross-wind shears, while still keeping the model simple and smooth.

The exact wind direction as well as the wake expansion parameter are varied to find the values that for the considered event minimizes the sum of absolute power deviations normalized by the total production:

$$\varepsilon = \frac{\sum_i \left|P_i - P_i^{\text{fit}}\right|}{\sum_i P_i} . \tag{5}$$

This choice of objective function for the fit does not allow over- and underestimations to cancel and is choosen in order to get a measure of how well the production pattern and not just total production is modelled.

The procedure described above uses simultaneous data from the turbines, i.e. a fixed (Eulerian) temporal reference frame. However, any change in wind conditions will need a finite time to propagate down the farm. We therefore also apply a co-moving (Lagrangian) reference frame, where the data from downwind turbines is taken with a delay estimated from the free wind speed. Denoting the the delay of the i'th turbine by τ_i and its downwind distance from the most upwind turbine by x_i, we use

$$\tau_i = \eta \, \frac{x_i}{v_{0,i}}, \tag{6}$$

where η is an empirical correction factor that we vary in order to find the overall optimum for our data set. This turns out to be at $\eta = 0.8$ and defines our co-moving reference frame (for details, see [9]). Note that we employ the heterogeneous free wind, i.e. the delay can vary also in the cross-wind direction.

3 Results

We find the optimal model parameters for each single event and can then investigate how these depend on various conditions. For example, in the right column of Fig. 1 is shown a scatter plot of the wake expansion parameter against the fitted wind direction for each of the three models. In all models the expansion parameter is higher for particular wind directions. These directions turn out to correspond nicely with major and minor symmetry axes of the Nysted wind farm layout. This finding is consistent with results from Ref. [4] which used a different Nysted data set. It should be noted that on average, the fitted wake expansion parameters are significantly lower than what is typically optimal when comparing to data averaged over many events: We find typical k values around 0.02 (as in Ref.[4]) and typical α values around 0.15–0.2.

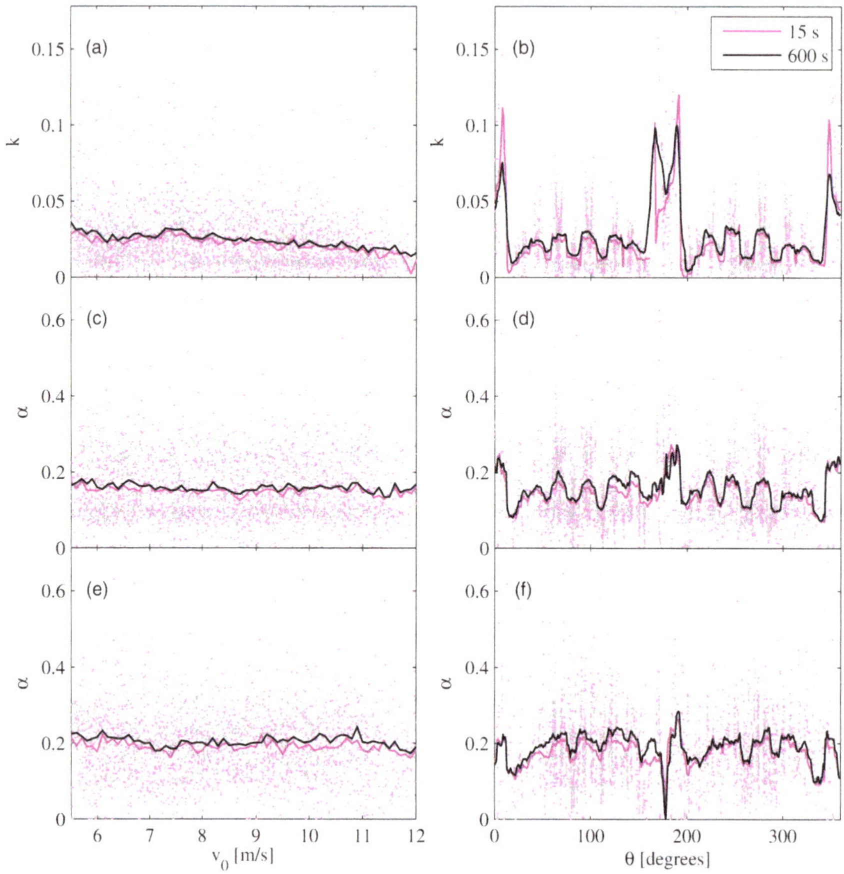

Fig. 1 Wake parameters of the Jensen model (top row) and the Frandsen model with Katic wake interaction (middle row) and semi-linear interaction (bottom row) as a function of the estimated up-wind speed (left column) and the fitted wind direction (right column). The dots represent all events with respect to 15 s (magenta) and 600 s (grey) averaging times. The lines are running averages based on a 0.1 m/s and 10° window width, respectively.

Averaging over a longer period will smoothen out many of the fluctuations due to turbulence and wake meandering that can not be captured by the simple wake models – we allow e.g. only a single wind direction for the whole farm. In Fig. 2 we plot the ensemble average fit error as a function of averaging time for the Frandsen model with Katic wake interaction. As a guide to the eye exponential curves with an offset are fitted to the points. The offset, i.e. the "asymptotic" error when the averaging time is increased can be seen as the limitation of the wake model even when short term fluctuations have been removed from the production pattern. Note that 10 minutes averaging puts one in the asymptotic regime. However, even then we are still left with a significant ensemble mean error of about 0.13 for the basic

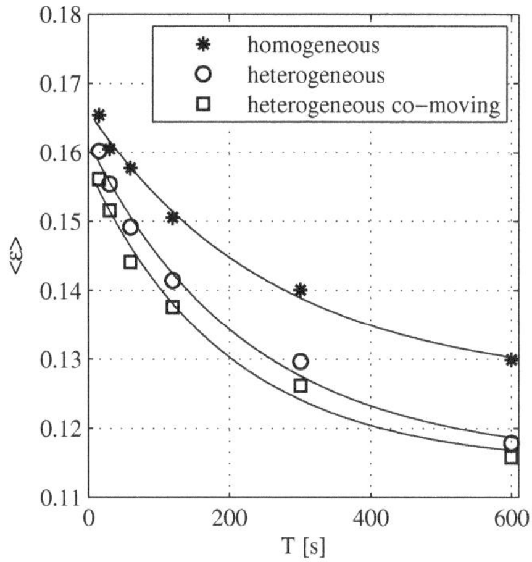

Fig. 2 Mean fit error of the Frandsen model with Katic wake interaction as a function of the event averaging time T for the data. The lines are fits of the form $a + b\exp(-T/T_0)$ with parameters $(a, b, T_0) = (0.126, 0.040, 260\text{s})$ (homogeneous), $(0.116, 0.046, 240\text{s})$ (heterogeneous) and $(0.115, 0.043, 200\text{s})$ (heterogeneous co-moving), respectively.

version of the Frandsen model with Katic wake interaction and homogeneous free wind. Using semi-linear wake interaction we find 0.15, while the Jensen model with Katic interaction gives 0.18 (not shown in Fig. 2).

In Fig. 2 we do not only report results from the basic model that assumes a single free wind speed for each event, we also evaluate the improvements that can be gained by the two refinements described in Sec. 2 above: Allowing heterogeneous wind speed in the crosswind direction (still keeping a single wind direction), and employing a co-moving temporal reference frame. As can be seen, allowing heterogeneous wind speeds delivers a significant improvement while also using the co-moving reference frame only slightly lowers average fit error. This is to be expected as only a fraction of the events are affected by propagating changes and for the majority of events, using a different reference frame will simply entail a new sampling of the random short-term fluctuations [9].

4 Conclusion

Existing engineering wake models have been applied in a single-event analysis of high-frequency wind-farm data. The tested models show comparable ability to describe the data. As expected the fit error increases with increasing time resolution – about 35% when going from a 10 min down to a 15 sec time resolution. The fitted wake expansion parameters turn out to be significantly smaller than usually obtained from standard sector-averaged model fits. Little dependence on the free wind speed was found for the Frandsen model, while the Jensen k parameter is found to decrease with increasing wind speed. For all model there was quite some variation of the average values with wind direction. This variation is not surprising given the highly simplified models and perhaps suggests a way to tweak them. Further studies are needed in order to quantify the optimal variation and to test whether it can used for predictions rather than just descriptions. An important generalization of the engineering wake models was the inclusion of heterogeneous cross-wind profiles, which lead to a significant improvement of the model fits. A smaller improvement was observed when a co-moving temporal reference frame was introduced in order to capture the propagation of changes in the wind conditions down the farm.

Acknowledgements. The authors acknowledge Dong Energy A/S and Siemens Wind Power for providing the data used in this study.

References

1. Ainslie, J.F.: Calculating the flowfield in the wake of wind turbines. Journal of Wind Engineering and Industrial Aerodynamics 27, 213 (1988)
2. Barthelmie, R.J., Jensen, L.E.: Evaluation of wind farm efficiency and wind turbine wakes at the nysted offshore wind farm. Wind Energy 13, 573 (2010)
3. Bjarnason, B., Sveinsson, Ö.: Wind farm controller - optimization of power production. Master's Thesis, DTU Electrical Engineering, Lyngby (2010)
4. Cleve, J., Greiner, M., Enevoldsen, P., Birkemose, B., Jensen, L.: Model-based analysis of wake-flow data in the Nysted offshore wind farm. Wind Energy 12, 125 (2009)
5. Frandsen, S., Barthelmie, R., Pryor, S., Rathmann, O., Larsen, S., Højstrup, J., Thøgersen, M.: Analytical modelling of wind speed deficit in large offshore wind farms. Wind Energy 9, 39 (2006)
6. Jensen, N.O.: Risø-m-2411: A note on wind turbine interaction. Risø National Laboratory, Roskilde (1983)
7. Katic, I., Højstrup, J., Jensen, N.: A simple model for cluster efficiency. In: Raguzzi, A. (ed.) European Wind Energy Association Conference and Exhibition, Rome, vol. 1, p. 407 (1987)
8. Machielse, L., Barth, S., Bot, E., Hendriks, H., Schepers, G.: Ecn-e–07-105: Evaluation of heat and flux farm control - final report. ECN, Petten (2007)
9. Poulsen, U.V., Scholz, J., Hedevang, E., Cleve, J., Greiner, M.: submitted to Wind Energy (2012)
10. Rathmann, O., Barthelmie, R., Frandsen, S.: Turbine wake model for wind resource software. In: Proceedings (online). 2006 European Wind Energy Conference and Exhibition, Athens, GR, 27 February- 2 March, European Wind Energy Association, Brussels (2006)

Turbulent Structures in Canopy Flows

Antonio Segalini, Jens H.M. Fransson, and P. Henrik Alfredsson

Abstract. A wind-tunnel experiment has been performed by means of a new forest canopy model. Hot-wire anemometry and Particle Image Velocimetry have been used to describe the flow from the statistical point of view together with a conditional sampling approach based on the Variable Interval Time Average method. The analysis demonstrated qualitative agreement between both measurement techniques, regardless of the canopy density that affects the structure intensity but not its topology.

1 Introduction

There is a renewed interest in the description of the atmospheric boundary layer over forested areas due to the strong exploitation of wind energy in Europe but also elsewhere. From the power production point of view, wind-turbine parks located on flat lands or offshore are more ideal as compared to locations in forested areas, but the number of such exploitable sites is limited. Therefore great efforts are spent to characterize the effect of the presence of a forest canopy in order to assess wind-turbine performance and the overall quality of a given site.

Wind-speed statistics profiles have been characterized by many experiments in both wind-tunnel [5, 6] and atmospheric boundary layers [2], providing useful fits of the mean velocity profile. However, there are still debates about the turbulent structures generated by the canopy-atmosphere interactions [8]. Poggi *et al.* [6] suggested that canopy flows show three main types of structures, namely the ones due to the vortex shedding inside the canopy, the ones due to the Kelvin-Helmholtz instability of the velocity profile (which has an inflectional point slightly above the canopy top) and finally the coherent structures typically found in wall-bounded turbulence.

Antonio Segalini · Jens H.M. Fransson · P. Henrik Alfredsson
Linné FLOW Centre, KTH Mechanics, SE-100 44 Stockholm, Sweden
e-mail: segalini@mech.kth.se

M. Hölling et al. (eds.), *Wind Energy – Impact of Turbulence*,
Research Topics in Wind Energy 2,
DOI: 10.1007/978-3-642-54696-9_13,

The distinction between these contributions is not an easy task even in laboratory settings, while it becomes impossible in atmospheric flows due to the limited spatial characterization of the flow field and the continuous changes of the geostrophic wind, which limits the identification of causes-effects of the atmospheric turbulence.

A wind-tunnel experiment has been therefore performed by means of a new forest canopy model. *Hot-wire anemometry* (HW) and *Particle Image Velocimetry* (PIV) have been used to describe the flow from the classical statistical point of view together with conditional sampling methods based on empirical techniques, often used in atmospheric flows [7]. Amongst the various existing methods, the VITA method (*Variable-Interval Time Average*) is here used to study localized variations in the velocity time series obtained from both HW and PIV over the wind tunnel forest model at some detection points [3, 7]. The use of two measurement techniques is expected to shed some light regarding the dynamics of the turbulent structures, since they are complementary techniques.

2 Experimental Setup

The experiments were performed in the Minimum Turbulence Level (MTL) wind tunnel at the Royal Institute of Technology (KTH) in Stockholm. The atmospheric boundary layer is simulated by means of triangular spires at the beginning of the test section to artificially increase the boundary-layer thickness up to 0.5 m. In order to simulate a forest canopy four pin fin plates with a width of 1.2 m and a total streamwise length of 2.0 m have been used. The canopy trees have been replicated by 5 mm diameter wooden circular pins that have been mounted in holes drilled in the plates, where the tip of the pins reaches $h_c = 50$ mm above the plate with a total model length of $40h_c$. The pins can be removed or added in order to create canopies with different densities. All the reported statistics have been measured at the streamwise station $x/h_c \approx 30$. As a notation, y will indicate a vertical coordinate with origin at the canopy top.

Hot-wire anemometry was used during the first part of the experimental campaign to determine statistical properties of the flow. Two HW sensors were used to educe the characteristics of the turbulent structures: A single HW probe (detector probe) and a crossed HW were used to measure time series that subsequently were phase averaged by using the detector probe. Time series have been acquired for five different free-stream velocities ($U_\infty \approx 5$, 10, 15, 20 and 25 m/s) at 20 points across the boundary layers with a sampling frequency of 20 kHz and a sampling time of 120 s in order to achieve good statistics and enough structures for the ensemble average to converge. Since the main statistics profiles did not show any Reynolds number dependence above the canopy (as shown in figure 1), only the phase averaged flow field with $U_\infty \approx 10$ m/s will be shown. During the two-point measurements, the used tree density of the canopy model was 850 pins/m^2, corresponding to an Element Area Index (EAI, defined as the frontal area per unit volume) of 17 m^{-1}.

For the PIV measurements one high-speed C-MOS camera (Fastcam APX RS, 3000 fps at full resolution, 1024×1024 pixels, Photron) was positioned in backward-forward scattering mode at an angle of approximately 90° to the acquisition plane. The raw images from the measurements had a resolution of 1024×1024 pixels and a 10-bit dynamic range. The images were taken at a sampling frequency of 800 Hz. A vertical laser light sheet of 1 mm thickness (aligned with the streamwise direction) was produced by a Nd-YLF laser (Pegasus PIV-Laser). Measurements were made at $U_\infty \approx 5$ m/s and $U_\infty \approx 10$ m/s.The canopy density during the PIV measurements was twice as high (1700 pins/m^2 with EAI= 34m^{-1}), with respect to the one used during the HW measurements, with the aim to quantify the canopy density effect on the educed structure. It is worth to mention that highest canopy density has been previously characterized with HW anemometry and no significant differences were observed in the mean velocity profile and in the Reynolds stress tensor above the canopy, with the exception of the vertical velocity variance, that increases with the increase of EAI.

3 Results

The mean velocity profile and velocity variances and co-variance in the vertical plane are shown in figure 1 for both HW and PIV for all the available U_∞. The friction velocity, u_*, is here used to normalize the velocity statistics and it has been determined from the maximum of the shear stress $-\langle u'v' \rangle$ measured during the HW measurement campaign. The mean velocity profiles from both measurement techniques show quantitative agreement, while the streamwise and wall-normal variances are in reasonable agreement. There are some discrepancies close to the forest probably due to excessive laser light reflection from the ground, or to the limitation of HW anemometry to measure recirculating flows, likely to occur at the canopy top. The statistics for both HW and PIV show that above the forest a high velocity shear is present, together with high turbulence intensity up to approximately $2h_c$. Above $2h_c$ the turbulence starts to decrease with height more rapidly than in the atmosphere due to the lack of free-stream turbulence in the present setup (besides the one provided by the spires), but it clearly points out that at $y \approx 2h_c$ there is the edge of the internal boundary layer due to the presence of the canopy.

The application of the VITA technique [7] to the HW data set allows the detection of time instants where significant variations are observed in the time series of the detection probe. Once these instants have been identified, a phase average can be performed and the associated characteristic structure identified. However, the structure signature will be defined in time, since only single-point time series are available with HW anemometry. Therefore the so called Taylor hypothesis must be applied as $\Delta x(y) = -U(y)\Delta t$, leading to the structure reported in figure 2, which is the phase average of the fluctuating velocity associated to the detection times. The picture of the educed structure is in agreement with atmospheric measurements over forests [1, 7], where large scale structures, educed with such empirical methods, are usually associated to some ejection of low momentum fluid from within the canopy, followed by a strong sweep of high momentum fluid from aloft.

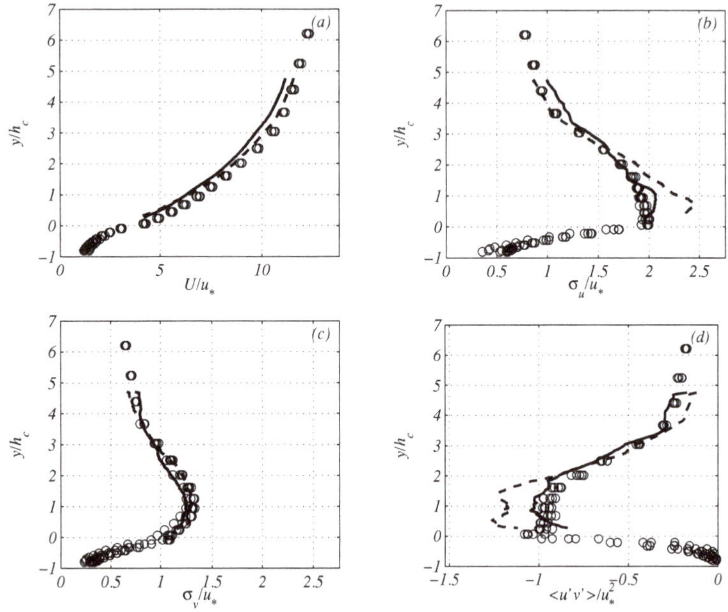

Fig. 1 Comparison of measured statistics profiles with HW with $5\text{m/s} \leq U_\infty \leq 27\text{m/s}$ (Circles), PIV with $U_\infty \approx 5\text{m/s}$ (Solid lines) and PIV with $U_\infty \approx 10\text{m/s}$ (Dashed lines). Mean streamwise velocity (a), standard deviation of the streamwise (b) and wall-normal (c) velocity and Reynolds shear stress (d).

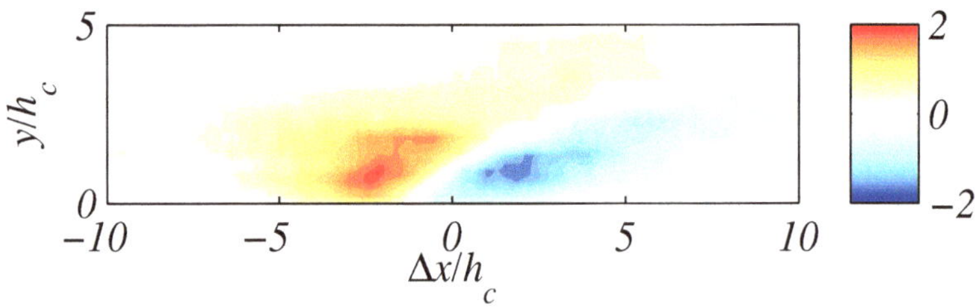

Fig. 2 Conditionally averaged streamwise velocity fluctuation field, $\Delta u/u_*$, detected by means of the VITA technique at $U_\infty \approx 10$ m/s (HW data). The used averaging time is $T_{av} = 0.7h_c/u_*$ with a threshold $k = 1$ [7].

The PIV measurements do not need the use of the Taylor hypothesis and can directly provide a spatial description of the structure. Figure 3 reports the PIV phase averaged structure at different instants relatively to the detection times, providing a slightly different description than the one reported in figure 2. It is possible however to observe that the same ejection-sweep cycle previously described is present but with a stronger structure intensity (probably associated to the different canopy densities).

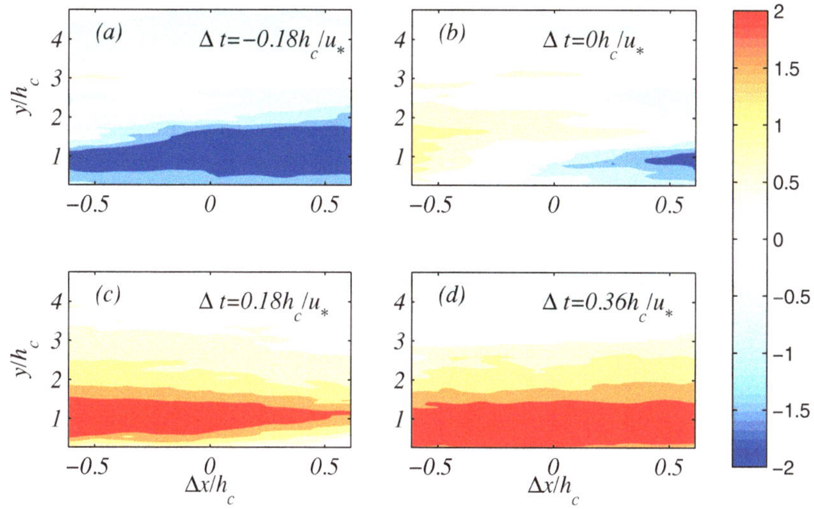

Fig. 3 Time evolution of the VITA structure detected by means of PIV at $U_\infty \approx 10$ m/s. (a)-(d) are subsequent phase averaged velocity fluctuation fields, $\Delta u/u_*$, with $\Delta t u_*/h_c = 0.18$. The used averaging time is $T_{av} = 0.7h_c/u_*$ with a threshold $k = 1$ [7].

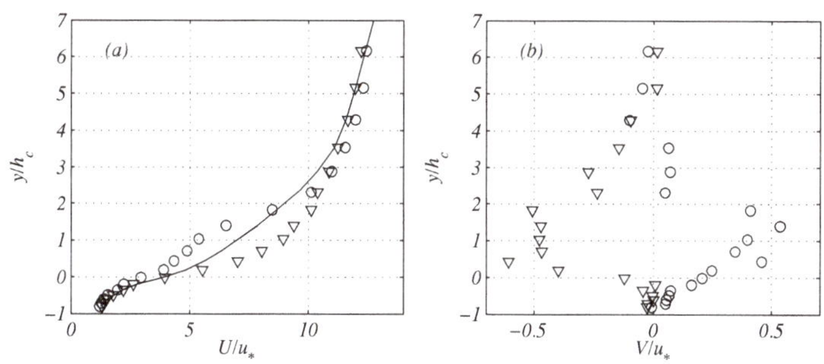

Fig. 4 Effect of the passage of the VITA event on the stream-wise and vertical velocity at $U_\infty \approx 10$ m/s (HW data). (▽) Sweep event, (○) Ejection event.

The effects of the structure passage on the mean velocity profile are more intense close to the canopy, where the highest velocity variation is observed, but they can also be observed even at $y/h_c \approx 5$ in weaker form, according to ref. [7]. Figure 4 shows the effects of such structure on the streamwise and vertical velocity profile. The ejection-sweep cycle is evident and it gives significant variations to the instantaneous wind velocity in a short amount of time, a phenomenon that might increase fatigue loads, as discussed by Odemark [4].

4 Conclusions

The effect of a forest canopy on the atmospheric boundary layer has been analyzed in the present work by means of simple statistical tools and phase averaged maps. Velocity statistics have been provided from both hot-wire anemometry and PIV, showing a good agreement in the mean velocity profile and a reasonable agreement in the covariances.

The VITA technique has been subsequently used to determine the instants where footprints of large scale passing structures were present, identified by means of a sudden variation in the velocity signal. The use of two independent measurement techniques allowed the comparison of the same eduction method by using measurement techniques with different time and space resolutions, together with an assessment of the outcomes of such empirical techniques, often applied to single-point statistics. Despite the fact that the VITA eduction method is not based on any velocity spatial gradient information, used in more sophisticated eduction criteria, its outcomes are noteworthy because they are associated to sudden streamwise velocity variations, events that have important implications in both wind-turbine energy production and fatigue loads.

The comparison of the structures educed with both techniques showed a qualitative agreement, demonstrating that the educed event is composed by an ejection-sweep cycle and it is convected downstream with approximately zero vertical velocity. However its intensity seems to increase with the increase of the canopy density, a property that clearly underlines the link between the educed structure and the canopy boundary condition. The effects of the passage of the educed structure on the mean wind speed profile have been demonstrated, pointing out that such large scale structures can enhance fatigue loads and provide challenges to pitch control systems of modern wind turbines located over forests.

Acknowledgements. *The Swedish Energy Agency*, under its program *Vindforsk III*, is acknowledged for the financial support.

References

1. Bergström, H., Högström, U.: Turbulent exchange above a pine forest. II. Organized structures. Bound-Lay Meteorol 49, 231–263 (1989)
2. Högström, U., Bergström, H., Smedman, A.-S., Halldin, S., Lindroth, A.: Turbulent exchange above a pine forest, I: Fluxes and gradients. Bound-Lay Meteorol 49, 197–217 (1989)
3. Johansson, A.V., Alfredsson, P.H.: On the structure of turbulent channel flow. J. Fluid Mech. 122, 295–314 (1982)
4. Odemark, Y.: Wakes behind wind turbines - Studies on tip vortex evolution and stability. TRITA-MEK technical report 2012:06, Stockholm (2012)
5. Raupach, M.R., Finnigan, J.J., Brunet, Y.: Coherent eddies and turbulence in vegetation canopies: the mixing-layer analogy. Bound-Lay Meteorol 78, 351–382 (1996)

6. Poggi, D., Porporato, A., Ridolfi, L., Albertson, J.D., Katul, G.G.: The effect of vegetation density on canopy sub-layer turbulence. Bound-Lay Meteorol 111, 565–587 (2004)
7. Segalini, A., Alfredsson, P.H.: Techniques for the Eduction of Coherent Structures from Flow Measurements in the Atmospheric Boundary Layer. Bound-Lay Meteorol (2012), doi:10.1007/s10546-012-9708-7
8. Shaw, R.H., Schumann, U.: Large-eddy simulation of turbulent flow above and within a forest. Bound-Lay Meteorol 61, 47–64 (1992)

A Closer Look at Second-Moment Closure for Wind Farm Analysis

Jonathon Sumner, Daniel Cabezón, Christian Masson, and Antonio Crespo

1 Introduction

It has been shown that common two-equation RANS turbulence closures are not well adapted to actuator disk modeling of the wind turbine rotor [9]. This combination of models is nonetheless incredibly popular for wind farm analyses given its relatively low computational cost and reasonably accurate velocity predictions for isolated rotors (see e.g. [1, 11, 7]). However, a fundamental change is likely needed to make significant improvements in velocity and turbulence predictions within a real wind farm. As the actuator disk representation of the rotor is by far the most economical in terms of computing time (in comparison to actuator surface or line approaches) and is probably a mainstay, it is more appropriate to reconsider the closure. Of course, the greatest weakness of many lower-order turbulence models is the reliance on the eddy viscosity approximation which ties stresses to strain rates in the flow and imposes a single turbulent length scale. On the other hand, Reynolds stress transport models avoid these problems entirely by modeling each component of the stress tensor with its own transport equation and might be a promising alternative.

Traditionally, stress transport models have received relatively less attention as the closure problem still has to be dealt with and the additional equations add to the computational burden. Furthermore, some question remains as to whether the extra

Jonathon Sumner · Christian Masson
École de technologie supérieure, Montreal, Canada
e-mail: jonathon.r.e.sumner@gmail.com,
christian.masson@etsmtl.ca

Daniel Cabezón
Centro nacional de energias renovables, Sarriguren, Spain
e-mail: dcabezon@cener.com

Antonio Crespo
Universidad polytechnica de Madrid, Madrid, Spain
e-mail: christian.masson@etsmtl.ca

M. Hölling et al. (eds.), *Wind Energy – Impact of Turbulence*,
Research Topics in Wind Energy 2,
DOI: 10.1007/978-3-642-54696-9_14,

effort in fact yields more accurate predictions. But, in light of the possible benefits, second-moment closure might be worth a closer look and is investigated herein for the purposes of wind farm power performance analysis. This study presents a comparison of measured and predicted power ratios for a wind farm sited in moderately complex terrain using two common two-equation closures and a stress transport model. In addition, two actuator disk implementations are investigated.

2 Mathematical Modeling

The objective of this study is to take a closer look at two important aspects of wind farm modeling in the context of popular RANS–actuator disk models: the turbulence closure and the actuator disk implementation. In general, the flow is considered steady and incompressible with the effect of turbulence on the transport of momentum being modeled with the Reynolds stress tensor and the effect of wind turbine rotors being represented by sink terms in the momentum equation.

Commonly, the turbulence closure is based on a two-equation parametrization of turbulence effects: the energy contained in the fluctuations and a measure of its rate of dissipation. These models are founded on the assumption that the momentum transport due to turbulence can be seen as stemming from an additional viscosity, generally orders of magnitude greater than the molecular viscosity, that may be deduced from local flow properties. The most popular two-equation RANS closure is the $k-\varepsilon$ model [6] which is well known to provide poor estimates of wind turbine wake properties. Although modifications to the original model have been proposed [4, 3, 8], a recent critique of eddy-viscosity-based models for wind turbine applications has shown that all models based on this formulation are probably weak [9]. These arguments provide the impetus to consider second-moment closures. Here, the $k-\varepsilon$ model, its RNG variant [12, 13], and the Reynolds stress transport model (RSTM) of Gibson and Launder [5] are tested. The model closure coefficients are calibrated for neutral surface layer flow where the von Karman constant, κ, is taken as 0.4187.

Concerning rotor modeling, the actuator disk method has been shown to yield nearly identical predictions of velocity defect as its higher-order counterpart the actuator line outside the very near wake region [10]. It is thus expected to be an adequate representation of the rotor for the purposes of evaluating wind farm energy capture as only far wake wind speeds are of interest. While coupling the actuator disk model with blade-element/momentum theory is quite feasible, it is somewhat impractical (as global thrust and power curves are often the only information available) and is likely unnecessary for such large-scale analyses.

However, application of actuator disk modeling in complex terrain and in densely-packed wind farms poses a unique problem when the actuator disk implementation is based on thrust and power curves. The challenge lies in the proper estimation of the reference wind speed. In theory, this is the wind speed at the location of the wind turbine rotor that would be observed if the wind turbine was not present: it represents the quantity of energy available at the rotor location.

Two possibilities are immediately apparent. The first relies on establishing a relationship between the wind speed at some 'freestream' location and the reference wind speed at the rotor. In its simplest form, a one-to-one ratio might be supposed. A more evolved approach might invoke some assumption regarding axial induction at the rotor itself; so far, improvements yielded by such methods for wind farms in complex terrain have been modest [8]. The first actuator disk implementation tested here follows this track and takes the wind speed two diameters directly upstream of the rotor as the reference wind speed. As the rotor thrust is determined iteratively as part of the flow solution it is referred to as an *elliptic* implementation.

The second possibility is to introduce the wind turbines sequentially in a row-by-row fashion such that the reference wind speeds for downstream turbines are calculated before their rotors are introduced. The advantage of such an approach is that the impacts of topography and wake recovery should be better captured although at the expense of longer run times. In this case, the rotor thrust is static and based only on upstream influences; it is thus referred to as a *parabolic* implementation.

In both cases, the elemental thrust introduced in the discretized momentum equations as a sink term is given by

$$\Delta \mathbf{T_i} = \frac{1}{2}\rho C_T\left(\|\hat{n}\cdot\mathbf{U}_\varnothing\|\right)U_\varnothing^2 A_\varnothing \frac{\Delta V_i}{V_\varnothing}\left(\hat{n}\cdot\hat{U}_\varnothing\right)^2\hat{n} \tag{1}$$

while the power is estimated using

$$P = \frac{1}{2}\rho C_P\left(\|\hat{n}\cdot\mathbf{U}_\varnothing\|\right)U_\varnothing^3 A_\varnothing \|\hat{n}\cdot\hat{U}_\varnothing\|^3 \tag{2}$$

where ρ is the air density, C_P and C_T are the turbine power and thrust coefficients, $\hat{n}$ is the disk-normal direction, $A_\varnothing$ is the swept area, $\Delta V_i/V_\varnothing$ is the fraction of the total rotor volume occupied by cell i, and $\mathbf{U}_\varnothing$ is the reference wind velocity.

There remains the question of the best measure of energy content. An argument could be made that an average based on the wind speed cubed would be most representative. In standardized power performance testing however the hub height wind speed is presumed to be representative. For wake flow, this practice is overly conservative: hub height wind speeds at the wake centre are either at or near minimum. As a first-order improvement, the disk-averaged wind speed is used as the reference wind speed $U_\varnothing$.

3 Case Study

The case study is based on a Spanish wind farm consisting of 43 wind turbines arranged in five rows with hub heights of 45 m and 55 m. The turbine thrust and power coefficient curves have been provided by the manufacturer. The terrain surrounding the wind farm is shown in Figure 1. A roughness length of $z_0 = 0.0082$ m is assumed for the entire site.

Observations have been filtered by sector and power as part of the European UPWIND project. The conditions for the present study correspond to a wind direction of $327° \pm 5°$, a wind speed at the reference turbine of 8.0 ± 0.5 m/s (by inverse power curve), and a turbulence intensity of roughly 12% at the meteorological mast at hub height.

4 Numerics

All simulations are run using OpenFOAM 1.7. The SIMPLE algorithm is used for pressure-velocity coupling. The convection terms in the momentum equation are discretized using a bounded version of QUICK; other convection terms use simple upwinding. All other terms are discretized with central differences. The simulations are run in parallel based on non-overlapping domain decomposition. Customized OpenFOAM solvers have been developed to introduce rotors "parabolically" and define reference wind speeds by the methods described above.

For the row-by-row introduction of rotors, the row convergence criteria are based on three parameters: the momentum equation residuals ($< 10^{-4}$), the maximum relative change in reference wind speed ($< 10^{-5}$), and a minimum number of iterations to ensure convection of upstream changes (500). For the elliptic actuator disk implementation, final convergence is based solely on normalized equation residuals; a tolerance of 10^{-5} is specified.

Equilibrium neutral surface-layer profiles of velocity, turbulent kinetic energy, dissipation rate, and stress tensor are specified at the inflow, located 11 km upstream of the wind farm. The friction velocity is calibrated to yield the measured power at the reference turbine. The outflow is placed 4 km downstream where a fully developed condition is assumed. The lateral boundaries, 2 km to each side of the wind farm, assume a symmetry condition. Freestream conditions are imposed at the upper boundary, roughly 1.5 km from the surface. Standard wall functions, with appropriate modifications for atmospheric flow [2], are used at the surface.

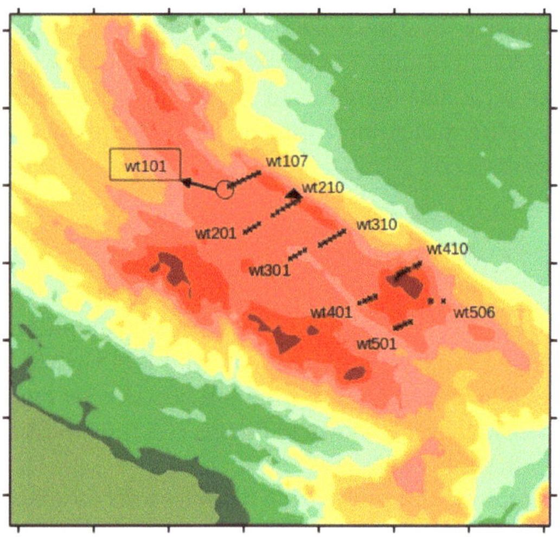

Fig. 1 Orography and layout of wind farm. Each contour level represents a 10-m change in elevation. Wind turbines are indicated by a cross while the meteorological mast is represented by a triangle. The reference turbine is wt101.

The domain is discretized using a structured mesh consisting of eight million cells with a near-wall cell height of 1 m; cell heights are expanded away from the wall. Each rotor is subdivided into roughly 40 cells.

5 Results and Conclusions

Figures 2 and 3 present the power ratios for the second and third rows of the wind farm. For the second row, the predictions of all models for nearly every turbine lie within one standard deviation of the observed power ratio. Significantly, no single combination of actuator disk implementation and turbulence closure stands out from the rest although the RSTM variants are generally somewhat more reliable than the two-equation closures. For row three, the agreement with measurement is not as impressive: all models over-predict the power production at turbine 7 with outliers also at 2, 3 and 5. Overall, the parabolic / RSTM model seems to provide the best agreement with measurements for this row, but only marginally so.

Returning to the original question regarding accuracy and computational effort related to the use of a stress transport model, the preliminary findings of this case study demonstrate that while some improvement in power predictions can be expected, it comes at a significant price: the run time (in terms of cpu-hours) for parabolic / RSTM was more than double that of parabolic / $k-\varepsilon$.

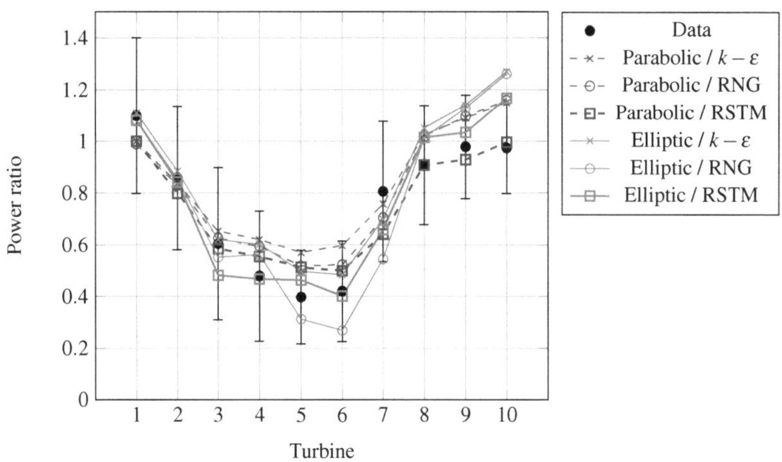

Fig. 2 Power ratios for turbines in second row. All results and measurements are normalized using the power output of the reference turbine (wt101).

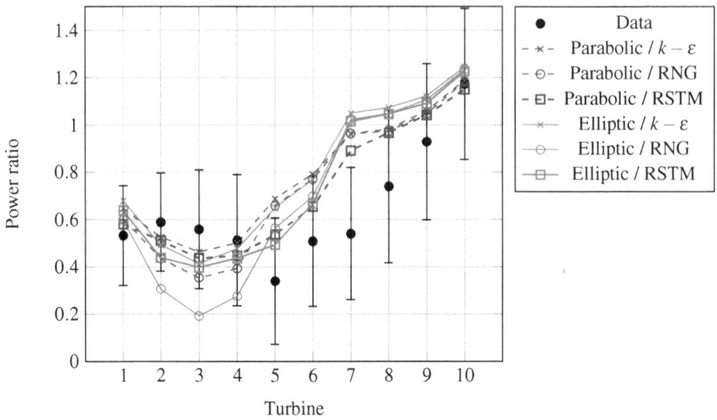

Fig. 3 Power ratios for turbines in third row. All results and measurements are normalized using the power output of the reference turbine (wt101).

References

1. Barthelmie, R.J., Hansen, K., Frandsen, S.T., Rathmann, O., Schepers, J.G., Schlez, W., Phillips, J., Rados, K., Zervos, A., Politis, E.S., Chaviaropoulos, P.K.: Modelling and measuring flow and wind turbine wakes in large wind farms offshore. Wind Energy 12(5), 431–444 (2009)
2. Blocken, B., Stathopoulos, T., Carmeliet, J.: CFD simulation of the atmospheric boundary layer: wall function problems. Atmospheric Environment 41, 238–252 (2007)
3. Cabezón, D., Migoya, E., Crespo, A.: Comparison of turbulence models for the computational fluid dynamics simulation of wind turbine wakes in the atmospheric boundary layer. Wind Energy 14 (2011)
4. El Kasmi, A., Masson, C.: An extended $k-\varepsilon$ model for turbulent flow through horizontal axis wind turbines. J. Wind Eng. Ind. Aerodyn. 96 (2008)
5. Gibson, M.M., Launder, B.E.: Ground effects on pressure fluctuations in the atmospheric boundary layer. Journal of Fluid Mechanics 86 (1978)
6. Jones, W., Launder, B.: The prediction of laminarization with a two-equation model of turbulence. Int. J. Heat Mass Transfer 15, 301–314 (1972)
7. Politis, E., Prospathopoulos, J., Cabezón, D., Hansen, K.S., Chaviaropoulos, P.K., Barthelmie, R.J.: Modelling wake effects in large wind farms in complex terrain: the problem, the methods and the issues. Wind Energy 15 (2012)
8. Prospathopoulos, J.M., Politis, E.S., Rados, K., Chaviaropoulos, P.K.: Evaluation of the effects of turbulence model enhancements on wind turbine wake predictions. Wind Energy 14 (2011)
9. Réthoré, P.-E.: Wind Turbine Wake in Atmospheric Turbulence. PhD thesis, Aalborg University (2009)
10. Réthoré, P.-E., Troldborg, N., Zahle, F., Sørensen, N.N.: Comparison of the near wake of different kinds of wind turbine CFD models. In: Proceedings of Gotland Wake Conference, Visby, Sweden (2011)

11. Sanderse, B., van der Pijl, S.P., Koren, B.: Review of computational fluid dynamics for wind turbine wake aerodynamics. Wind Energy 14 (2011)
12. Yakhot, V., Orszag, S.A.: Renormalization group analysis of turbulence. I. Basic theory. Journal of Scientific Computing 1(1), 3–51 (1986)
13. Yakhot, V., Smith, L.M.: The renormalization group, the ε-expansion and derivation of turbulence models. Journal of Scientific Computing 7(1), 35–61 (1992)

Part III
Experiments

Small Scale Sensing for Wind Turbine Active Control System

Rustom B. Bhiladvala*, Elsa Assadian, and Ali Etrati

Abstract. In large wind-turbine parks (farms), gusts, turbulence and interaction of wakes cause significant variation in the wind velocity over rotor blades, with a range of temporal and spatial scales. Arrays of sensors for pressure and wall shear stress at a few key locations, would directly provide data on local wind forces at these locations on the blade surface. Augmented by wind speed field data, wall shear data would be the most directly relevant input for "smart rotor" Active Flow Control (AFC), reducing power intermittency due to localized or large-area boundary-layer separation. Actuators run by a control system with such input could reduce extreme and fatigue loads, power fluctuations, and could help extend the wind speed envelope for power extraction. For measuring fluctuations of turbulent wall shear stress, flush-mounted single-hot-film sensors are non-intrusive, but suffer from unacceptably large errors caused by significant heat conduction to the substrate. To mitigate this problem, we introduce multi-element guard-heated sensors. Results from our analytical/numerical study show that it is possible to eliminate substrate conduction errors, with careful geometric design of the guard heater. A microfabricated prototype is also presented in this entry.

1 Introduction: Sensors, Actuators, Control for Wind Energy

Large wind turbines (WTs) in a wind park typically experience significant force fluctuations, over a range of time-scales (or frequencies). The timing and severity of wind turbine component failures from fatigue and extreme loads increases the cost of electricity produced and more importantly, could damage the perception of wind as a dependable (annually-averaged) renewable energy resource. This dependability is required for citizens, governments and markets to support its development.

Rustom B. Bhiladvala · Elsa Assadian · Ali Etrati
IESVic & Department of Mechanical Engineering, University of Victoria, Victoria BC, Canada
e-mail: {elsa,rustomb}@uvic.ca

* Corresponding author.

M. Hölling et al. (eds.), *Wind Energy – Impact of Turbulence*, Research Topics in Wind Energy 2,
DOI: 10.1007/978-3-642-54696-9_15,

A coordinated strategy for sustained wind energy development with arrays of large wind turbines may therefore reasonably include the costs of:

(1) Active Flow Control (AFC) system -effective, low-maintenance, long-lasting.
(2) Distributed sensors for sampling local wall shear stress and pressure. Other sensors for wind speed, temperature, other weather conditions.
(3) Algorithms for Active Flow Control -robust, stable, optimized with WT, wind-park and grid-related variables.
(4) Processing of sensor data with [i] a higher resolution stream for smart rotor control system input - and [ii] coarse-grained data saved for WT health monitoring, improvements in design, controls, materials and manufacturing methods.
(5) Scheduling minimal pre-emptive maintenance based on WT health monitoring.
(6) Analytical, Computational and Experimental Modeling, crucial for(1)-(5).

The development of distributed sensors (item 2) for sampling local blade forces caused by wind is the subject of this paper. Currently, with increasing size of wind turbines, there appears to be growing interest in developing AFC, with the primary aim of reducing extreme and fatigue loads arising from wind variability [1]. Several possible forms of actuators (1) for smart rotor or smart blade AFC have been described [2, 3, 4] and progress reviewed [5]. Blade-strain sensors are in use [5] for WT health monitoring (4) and can serve mimimal maintenance goals (5).

LIDAR for wind velocity field sensing upto kilometre scale, can serve for WT feedforward control systems [5]. Distributed sensors (2) and control algorithms (3) for smart-rotor WT to grid systems appear not to have received as much attention [6]. In a recent study, Peinke and co-workers [7] concluded that wind fields parameterized by mean speed and fluctuation intensity, and generated by standard algorithms, do not adequately reproduce the intermittency of measured wind fields, and that statistics of small-scale features of wind turbulence need to be better understood and represented.

Temperature-compensated turbulent wall shear stress (wss) sensors would directly provide high resolution measurements of local wss caused by gust onset (rapid rate of change) as well as turbulent fluctuations(4). Measurements of local pressure and wss with appropriate resolution could be used as an input to an AFC system (3) to reduce local flow separation, as the wind-field changes (4). Such an AFC system would deploy actuators (1) to reduce extreme or fatigue loads caused by intermittent gusts or wakes (4). Potential benefits for AFC could include an extended wind-speed envelope for power extraction; control vortex shedding and noise generation; provide wind-force history for blades, by saving coarse-grained data; interpret strain or micro-damage history and weather history, to plan minimal maintenance (4,5); seek blade materials and manufacturing techniques for longer wind turbine life.

Controlling cost and complexity implies that the locations of small arrays of such sensors would have to be chosen judiciously, following some experimentation and modeling. Direct wind-force blade data from these sensor arrays on model blades in controlled flow, could be used to empirically enhance fast, inexpensive, inviscid aerodynamic models (6). The sensors being developed here will also enable laboratory turbulent boundary layer studies with simultaneous measurement of wall-shear and velocity fields, to improve wall-boundary condition specification for more compute-intensive modeling (6) with LES (large eddy simulation).

2 Guard-Heated Thermal Sensors for Turbulent Wall Shear Stress

Even in well-controlled laboratory turbulent flows with constant mean velocity, the probability density functions (pdfs) of wss show heavy tails with fluctuations far greater than the mean being significant. The rapid onset of wind gusts may contribute to an intermittency in power fluctuations, both through local separation as well as interaction with near-wall turbulence structures. Accurate measurement of wss fluctuations should help our understanding, and techniques used to do so must not appreciably perturb the sublayer, which becomes thinner as the mean skin friction increases. This is the reason why sensors or sensing volumes with dimensions far smaller than the typical large scales in the flow are required for wss measurement. The advent of microfabrication has seen several new small-scale techniques such as micropillar deflection and floating-element microsensors. For the

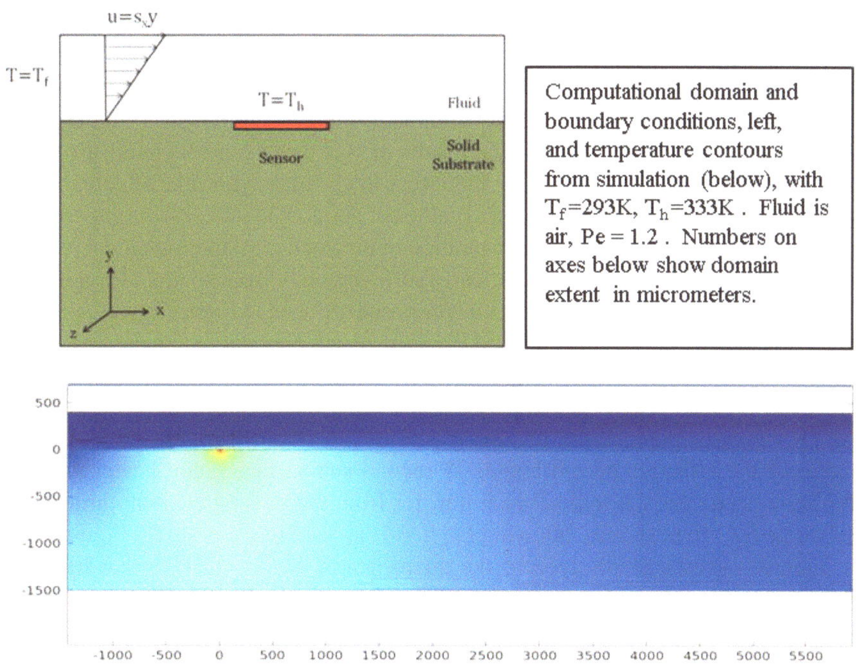

Fig. 1 Conjugate heat transfer problem. All external boundaries have zero temperature gradient, a flux-continuity condition is used at the fluid-solid interface, S_x is the shear rate, u(y) is the linear velocity profile. Temperature contours for a conventional single-element hot film in air show heat exchange over a far larger length than the hot-film length, seen in red.

operating environment of wind-turbines, flush-mounted thermal sensing films with a thin protective dielectric coating, seem to be the most rugged option likely to survive the environment.

In constant temperature anemometry (CTA) with a flush-mounted hot-film, the relation between measured voltage (E) and wss (τ) used is:

$$E^2 = A\tau^{1/3} + B, \tag{1}$$

where constants A and B are obtained from a calibration experiment, using a flow with known shear stress. This relation was first derived by Leveque [8], from the advection-diffusion equation for heat transport from an isothermal hot film to moving fluid using several assumptions. Analytical and computational studies of the conjugate fluid-solid heat transfer show that these assumptions, particularly negligible substrate conduction, can be grossly violated, particularly in the case of low thermal conductivity fluids such as air.

Figure 1 (top) shows the domain and boundary conditions used for the conjugate heat transfer problem. The advection-diffusion equation is solved for the fluid, heat conduction equation for the solid, and continuity of heat flux is used at the fluid-solid interface. Results for a single-element (SE) hot-film sensor in air (seen in Figure 1, bottom, and the SE curve in Figure 2) show the effective heated length of the fluid-substrate interface can be over a hundred times larger than the physical length (reducing spatial resolution), and changes significantly with shear strength (Pe). The Peclet number $Pe = \frac{S_x L^2}{\alpha_f}$ is the shear strain rate S_x non-dimensionalized by the fluid thermal diffusion timescale, based on hot-film length L and the fluid thermal diffusivity α_f. Experimental results of wss measurements in air [9] with a conventional single-element sensor show very low values of turbulent fluctuation intensity, from $0.06 - 0.12$, compared to values of about $0.34 - 0.37$ from experiments in liquids [9] and from DNS. The resolution of conventional single-element hot-film sensors for wall shear stress in air clearly suffers greatly from spatial averaging.

Multi-element guard-heated sensors proposed are seen in Figure 2, top. All guard heaters (blue) are electrically isolated from the sensor element, but maintained at the same temperature, using a single, separate, CTA bridge circuit. The single-element (SE) sensor (red), can be equipped with a single guard heater in-plane (GH1P), which increases the fraction of sensor heat directly transferred to the fluid (Q_f), reducing the effective heated length. A second sub-surface guard-heater is called for in air, to reduce the indirect heat transfer from the sensing element through the substrate (Q_s) to near-zero, allowing $L_e/L = 1$, as seen in the plots for Figure 2.

Figure 3 shows a prototype of the guard-heated sensor (GH1P) that has been fabricated. The chip is designed so that heat generated in the inner sensing element (the green rectangle seen on the chip) is forced by a surrounding guard heater (blue) maintained at the same temperature by a second bridge circuit, to be transferred directly to turbulent flow of air over the chip face. For the sensors with guard-heating in two planes, substrate conduction can be entirely eliminated, but the more difficult fabrication for this is currently being worked out.

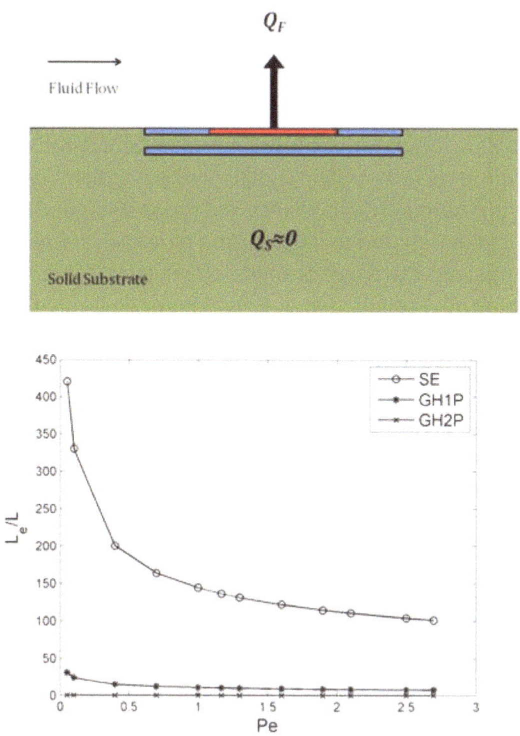

Fig. 2 Upper: Sketch defining single-element sensor SE (red-only), sensor with guard-heating in 1-plane, GH1P(red+ top-blue) and guard-heating in 2-planes, GH2P(GH1P + additional guard-heating film below sensor). Lower: Effective sensor (heat-exchange at interface) length L_e for SE can range from 150 to 200 times the sensor hot-film length L for Pe between 1 and 0.4; in air, it is still significant for the GH1P case, and exactly equal to L (ideal) for the GH2P case (see text).

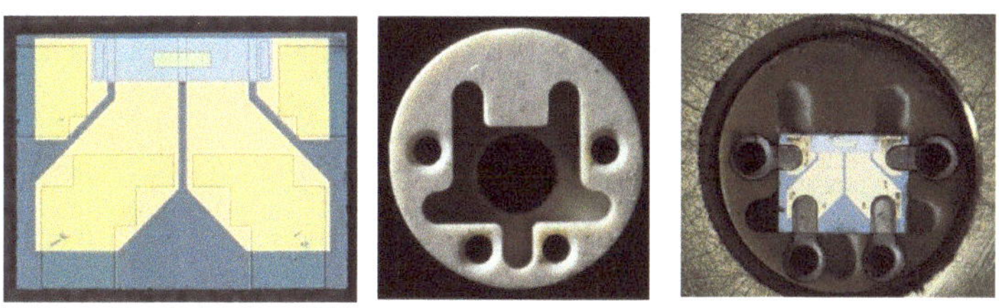

Fig. 3 Prototype of guard heated GH1P wall shear stress sensor: chip, ceramic holder and leads

3 Conclusions

A system of distributed sensors of local pressure and wall shear stress at key locations on wind turbine blades could be useful to directly measure the local wind force on blades, without the need for models that transform wind-field data to blade force data. They can be designed for the resolution required. Wall shear stress sensors can provide valuable data on incipient local flow separation, and may thus help to mitigate extreme and fatigue loads, vortex shedding, noise and stall through the deployment of actuators by an Active Flow Control or "smart-rotor" system. An economical low-resolution record of their signals will have many uses, which include health monitoring, pre-emptive maintenance and the analysis of aging or failed components to improve materials and manufacturing processes for new design. Though rugged and non-intrusive, conventional single-element hot films suffer from serious errors due to substrate heat conduction. Detailed numerical calculations for the proposed microsensors with guard heating indicate they should be successful in eliminating errors from substrate heat conduction. First prototypes have been fabricated.

References

1. Johnson, S.J., Baker, J.P., van Dam, C.P., Berg, D.: An Overview of Active load Control Techniques for Wind Turbines With an Emphasis on Microtabs. Wind Energ. 13, 239–253 (2010)
2. Sandia Report, SAND2008-4809, Sandia National Laboratories (2008)
3. Christian Bak, C., Gaunaa, M., Andersen, P.B., Buhl, T., Hansen, P., Clemmensen, K.: Wind tunnel test on airfoil Ris-B1-18 with an Active Trailing Edge Flap. Wind Energ. 13, 207–219 (2010)
4. Maldonado, V., Farnsworth, J., Gressick, W., Amitay, M.: Active control of flow separation and structural vibrations of wind turbine blades. Wind Energ. 13, 221–237 (2010)
5. Barlas, T.K., van Kuik, G.A.M.: Review of state of the art in smart rotor control research for wind turbines. Progress in Aerospace Sciences 46, 1–27 (2010)
6. van Kuik, G.A.M., Berg, D.E.: Editorial, Special Issue. Wind Energ. 13, 101 (2010)
7. Mücke, T., Kleinhans, D., Peinke, J.: Atmospheric turbulence and its influence on the alternating loads on wind turbines. Wind Energ 14, 301–316 (2011)
8. Leveque, M.: Leveque Leslois de la transmission de chaleur par convection. Annus Mines Ser. 12(13), 234–242 (1928)
9. Alfredsson, P.H., Johansson, A.V., Haritonidis, J.H., Eckelmann, H.: The Fluctuating Wall Shear Stress and the Velocity Field in the Viscous Sublayer. Physics of Fluids 31(50), 1026–1033 (1988)

Wind Tunnel Simulation of Wind Turbine Wakes in Neutral, Stable and Unstable Offshore Atmospheric Boundary Layers

Philip E. Hancock, Frauke Pascheke, and Shanying Zhang

Abstract. Relatively little is known about wind turbine wake development in stable and convective wind flow. From the UpWind [1] study wind conditions in areas of the North Sea are significantly non-neutral for about 70 percent of the time. The stable atmospheric boundary layer is particularly complex, and if the stability is sufficiently strong the boundary layer depth is comparable with the tip height of large wind turbines. The 'imposed' condition, the potential temperature gradient above the boundary layer, is therefore important. In a first phase of work, measurements have been made in the wake of a model wind turbine in neutral, stable and unstable offshore atmospheric boundary layer simulations, the neutral case being the reference case. The stable case is for weak surface-layer stability, but typical strong imposed stability above. The wake deficit decreases more slowly, consistent with the lower level of boundary layer turbulence, but after about 3 rotor diameters it ceases growing in height as a direct consequence of the imposed stability. Temperature and heat transfer are increased in the wake, the latter though only from about 3 diameters. The unstable cases show more rapid reduction of the velocity deficit, and higher levels of turbulence. The greater depth of the unstable layer means that conditions above are much less important.

1 Introduction

Understandably, a widely-made assumption in the wind power industry regarding wind resource is that the wind flow is neutrally stable or nearly so, and that unstable and stable periods roughly balance out and are equivalent to neutral conditions. (See for instance [2] regarding wake prediction.) However, the mean wind profile and the level of turbulence in the atmospheric boundary layer (ABL) are dependent

Philip E. Hancock · Frauke Pascheke · Shanying Zhang
University of Surrey, Guildford, United Kingdom
e-mail: p.hancock@surrey.ac.uk

M. Hölling et al. (eds.), *Wind Energy – Impact of Turbulence*,
Research Topics in Wind Energy 2,
DOI: 10.1007/978-3-642-54696-9_16,

upon stability, and wind flow turbulence has long been recognised as promoting wake diffusion and a more rapid deficit reduction (see for example [3]), potentially allowing a smaller streamwise separation between turbines. Turbulence and mean shear from an upwind turbine adds to the load imposed on a turbine over and above that of the turbulence and mean shear arising in the undisturbed wind flow. Modelling wakes is therefore of major practical importance. However, relatively, little is known quantitatively about how wakes develop under stable or unstable wind flow conditions.

The convective motions in an unstable ABL give rise to higher intensities and greater depth, while the buoyancy forces in the stable case reduce the intensities and depth. Importantly in this latter case, the depth of the boundary layer can be reduced enough for the top of it to be below the top of the rotor disk of a large wind turbine (say 5MW, where the top of the rotor disk is at about 150m). In the neutral and unstable cases the whole height of the turbine is immersed in the ABL. The stable ABL is particularly complex, and the condition imposed from above by the stable temperature gradient, $d\theta/dz$, is as much a principal parameter as those of the surface layer. This is an important point for wind power meteorology because there is little field data for temperature and velocity profiles up to heights of 200m.

The work reported here is part of a programme of work within the UK-EPSRC Supergen-Wind consortium. A first series of wind tunnel measurements have been made of a single turbine in stable and unstable boundary layers.

2 Wind Tunnel and Turbine Models

The EnFlo wind tunnel is designed to simulate neutral and stratified atmospheric boundary layers. Stable or unstable stratification is created by cooling or heating the floor surface, assisted by heaters upstream of the working section that can provide a temperature gradient at the inlet. A heat exchanger removes heat at the end of the working section. Irwin-type spires [4] at the working section inlet together with surface roughness are used to generate mean velocity and turbulence profiles with typically correct characteristics. Here, the neutral case was based on the guidelines of ESDU [5, 6], assuming a mean wind speed of 10m/s at a height of 10m. The working section is 20m long, 3.5m wide and 1.5m high, and the simulation was done at a scale of 300:1. The three-blade model turbine has a diameter of 416mm and a hub height of 300mm, with a design tip-speed-ratio of 6. The rotor drives a four-quadrant motor-generator through a gearbox. A consequence of scaling is that the temperature gradient $d\theta/dz$ above a stable boundary layer must be such that

$$\left(\frac{D}{U_{ref}}\right)^2 \frac{d\theta}{dz} = constant, \tag{1}$$

where U_{ref} is a reference velocity, D the rotor diameter and z the height from the surface. This implies that the gradient in the wind tunnel must be much larger than in the atmosphere; non-dimensionally, LN/U_{ref} will be the same for both for similarity, where L and N are the Obukhov length and the Brunt-Vaisala frequency, respectively.

Velocity was measured by means of a Dantec FibreFlow two-component frequency-shifted LDA probe, mean temperature by thermocouples and fluctuating temperature by a cold wire held close to the measuring volume in order to measure the instantaneous heat fluxes. The cold wire was calibrated against the traversed thermocouple. Each measurement point was sampled for a minimum of 2 minutes, at a mean sampling rate of about 90Hz for the LDA, and at 3kHz for the fluctuating temperature, interpolated to coincide with the LDA samples. The largest error is expected to be in the statistical averaging, taken to be within about $\pm1\%$ for mean velocity and $\pm10\%$ for the second-order moments. The technique, including details about the frequency response for the temperature fluctuations, is described in more detail by [7].

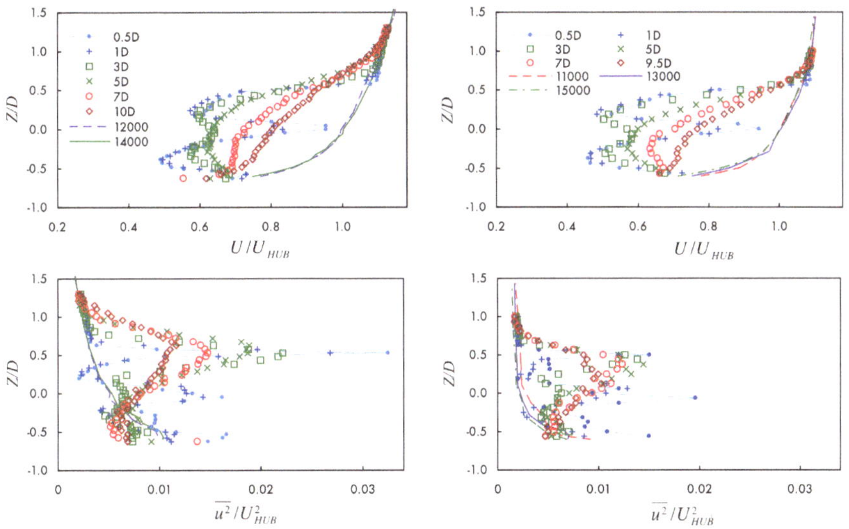

Fig. 1 Wake profiles of mean velocity and Reynolds direct stress $\overline{u^2}$ between 0.5D and 10D from the turbine. Left side, neutral, right side stable. Lines indicate respective undisturbed profiles (at distances from the working section inlet (mm)).

3 Results

Figure 1 shows profiles of the mean velocity U and streamwise Reynolds stress $\overline{u^2}$, normalised by the hub-height velocity, U_{HUB}. The left side of the figure is the neutral case and the right the stable case, for which L/D was 3.0. Here, z is from the hub axis. The lines in the figures show the undisturbed profiles at X from the working section inlet, in units of mm.

There are number of immediately notable features. The momentum deficit is larger for the stratified case, as is to be anticipated by the lower level of turbulence in the ABL. However, the vertical growth rate is also clearly less at the downstream stations, and in fact has ceased, as can be seen by the comparison given for the last two stations in figure 2 (a feature also seen in other quantifies). A closer examination of the profiles shows that the growth rates are very similar up to about 3 diameters, but is suppressed from 5D onwards. This is inferred as a direct effect of stratification, while the larger deficit in the earlier part of the wake is an indirect one, arising from a lower level of ABL turbulence. The Reynolds stress ($\overline{u^2}$) profiles are noticeably smaller in magnitude in terms of "added" turbulence of $\overline{u^2}$ - the undisturbed level. That the stress is smaller in magnitude for the earlier profiles is interesting in that the mean shear is larger in the stable case.

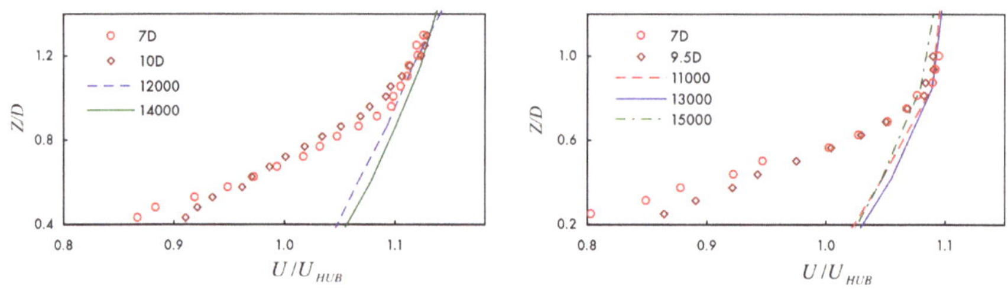

Fig. 2 Mean velocity profiles, near the profile edge, from figure 1

Also noticeable is the *lower* level of $\overline{u^2}$ w.r.t. the unperturbed ABL profile, but only in the neutral case. Tentatively, it is thought this may be due to an impeding effect of the turbine on axial fluctuations in the upstream flow, though [11] attributed it to a decrease in mean velocity gradient, but by either mechanism it should also be present in the stable case.

Qualitatively similar results to that seen here of a larger wake deficit were also seen in field measurements by Magnusson and Smedman [8], and opposite to wind tunnel experiments by Chamorro and Porte-Agel [9], contrary to expectation, for reasons that are not understood, though it may be linked with a smaller (but not zero) imposed condition in their case.

Figure 3 shows the mean temperature profiles and the heat flux profiles, where the latter are normalised by the level in the undisturbed ABL. As in figure 1, the lines in the figure indicate the undisturbed profiles. Above the wake, the temperature gradient is about 20K/m which, from equation (1), is equivalent to 0.01K/m at full scale. (*D* was in the ratio of 1:300, model-to-full scale, and $U_{ref} intheratioof$ 1.5:10.) This a typical strong gradient. See, for example, [10]. The imposed condition is therefore 'strong', but at the surface, from *L*, the stratification would be classed as 'weak'.

The wake has a significant effect on the mean temperature profile, giving a distinct increase below about hub height, at all stations. In contrast, the heat flux measurements show little initial response; it is only from 3*D* onwards that a dramatic

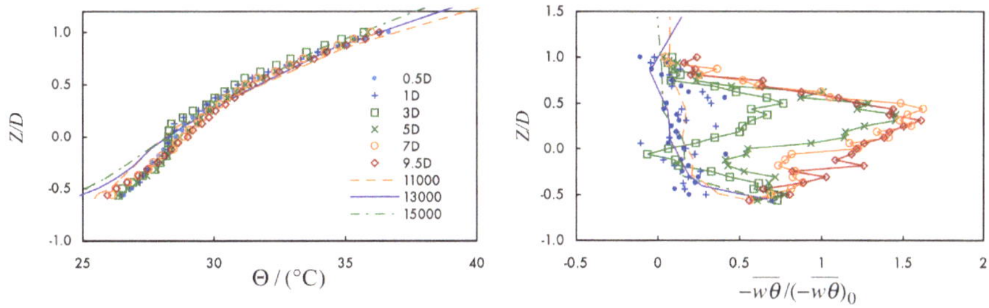

Fig. 3 Mean temperature profiles (left) and vertical heat flux profiles (right), the latter normalised by the unperturbed surface value

change takes place, with a peak level exceeding the surface value by a factor of about 1.6, and the steep change in heat flux, once at the surface, is now at the top edge of the wake. The change in response after about 3*D* is consistent with the observations above about development of the mean velocity profiles being affected directly by stratification only after about 3*D*.

Figure 4 shows contour plots of mean velocity and Reynolds shear stress $\overline{uw}$ for two unstable surface conditions, L/D. The mean velocity deficit can be seen to decrease more rapidly with decreasing $|L|$, and the shear stress becoming more intense. The shear stress contours also indicate a more rapid growth rate. These latter measurements are preliminary in that the undisturbed ABL was still evolving spatially, while for the stable case, as can be seen in figs 1 and 3 there was a negligible or small change. Simulation improvements are the subject of current work.

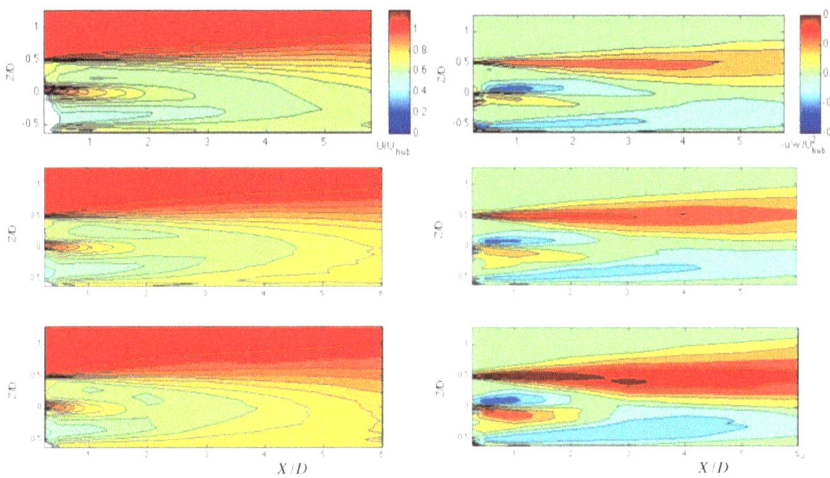

Fig. 4 Contours of mean velocity U and Reynolds shear stress $\overline{uw}$, normalised by U_{HUB}. Top pair, neutral; middle, $L/D = -2.8$; bottom, $L/D = -1.6$.

Acknowledgements. The authors wish to express their thanks to colleagues Tom Lawton, Dr Paul Hayden and Allan Wells for assistance in setting up the experiments, to Prof Alan Robins for his support and various discussions on wind tunnel simulation, etc, and to Prof Albert Holtslag for useful discussions at this EuroMech meeting on boundary layer characteristics, field measurements and modelling. The authors also wish to thank colleagues in the Supergen-Wind consortium (www.supergen-wind.org.uk) and EPSRC in supporting this work.

References

1. UpWind. Flow and wakes in large wind farms: Final report for UpWind WP8. Ris report R-1765 (EN) (2011) ISBN 978-87-550-3878-3
2. Sanderse, B., van der Pijl, S.P., Koren, B.: Review of computational fluid dynamics for wind turbine wake aerodynamics. Wind Energy 14, 799–819 (2011)
3. Ainslie, J.F.: Calculating the flow flied in the wake of wind turbines. J. Wind Eng. and Industrial Aerodynamics 27, 213–224 (1988)
4. Irwin, H.P.A.H.: The design of spires for wind stimulation. J. Wind Eng. and Industrial Aerodynamics 7, 361–366 (1981)
5. ESDU: Characteristics of atmospheric turbulence near the ground. Part II: Single point data for strong winds (neutral atmosphere). ESDU 85020. Engineering Sciences Data Unit (2001)
6. ESDU: Strong winds in the atmospheric boundary layer. Part 1: hourly-mean wind speeds. ESDU 85026. Engineering Sciences Data Unit (2002)
7. Heist, D.K., Castro, I.P.: Combined laser-doppler and cold wire anemometry for turbulent heat flux measurement. Exp. in Fluids 24, 375–381 (1998)
8. Magnusson, M., Smedman, A.-S.: Influence of atmospheric stability on wind turbine wakes. J. Wind Engineering 18, 139–152 (1994)
9. Chamorro, L.P., Port-Agel, F.: Effects of thermal stability and incoming boundary layer flow characteristics on wind turbine wakes: a wind tunnel study. Boundary Layer Met. 136, 515–533 (2010)
10. Steeneveld, G.J., van de Weil, B.J.H., Holtslag, A.A.M.: Diagnostic equations for stable boundary layer height: evaluation and dimensional analysis. J. Appl. Meteorol. and Climatol. 46, 212–225 (2007)
11. Hassan, U.: Wind tunnel investigation of the wake structure within small wind turbine farms. Energy Technology Support Unit report, ETSU WN 5113 (1993)

Unsteady Flow Investigation around a Pitching Wind Turbine Blade Element

Kobra Gharali, David A. Johnson, and Vivian Lam

Abstract. A numerical and experimental investigation is presented studying the unsteady wake and evaluating the unsteady aerodynamic loads for an S822 airfoil in pitching motion, for the case where the maximum dynamic angle of attack is equal to the static angle of maximum lift to drag ratio. Smoke wire flow visualization compares the wakes of static and dynamic cases for $Re < 10^5$ and shows the strong dependency of the wake structure on the Reynolds number. Laser Doppler Anemometry (LDA) measures the unsteady wake characteristics of the flow for $Re = 10^5$ and shows that increasing the amplitude of oscillation increases the boundary layer thickness. For all cases, a numerical simulation with the Transition SST method has been employed as an alternative method and shows good agreement with the experimental results. Finally, the unsteady loading will cause varying rotor loads.

1 Introduction

Wind turbines are renewable energy devices that operate in variable and unsteady environments. In some situations, the turbine rotors are yawed which results in unsteady airloads [1]. Studying yawed wind turbine rotors are important even for pitch controlled [2] and small horizontal axis wind turbines (HAWT) [3, 4, 5, 6]. A yawed rotor has a periodic flow field structure that can be modeled by a 2D pitch oscillating airfoil. Oscillating airfoils with application in HAWTs have been studied numerically and experimentally to simulate yawed rotors, but most consider high Reynolds number and there are fewer studies for low Reynolds numbers ($Re \approx 10^5$) when laminar-turbulent transition plays a significant role and challenges numerical methods. In this study a S822 airfoil, designed for small HAWTs [7], has been

Kobra Gharali · David A. Johnson · Vivian Lam
Wind Energy Group, Department of Mechanical and Mechatronics Engineering, University of Waterloo, Ontario, Canada
e-mail: {kgharali,da3johns,v5lam}@uwaterloo.ca

M. Hölling et al. (eds.), *Wind Energy – Impact of Turbulence*, Research Topics in Wind Energy 2,
DOI: 10.1007/978-3-642-54696-9_17,

chosen. Based on the authors' knowledge, there are no publications studying the flow field of the dynamic S822 airfoil. The airfoil oscillates sinusoidally, according to $\alpha = \alpha_{mean} + \alpha_{amp}\ \sin(2\pi f t)$, where mean angle of attack (α_{mean}), pitch oscillation amplitude (α_{amp}), and oscillation frequency (f) define an oscillation system with the reduced frequency k ($k = \frac{\pi f c}{U_\infty}$, c is the chord length). In this study, the experimental and numerical results are presented for $\alpha_{mean} = (0^o)$, $\alpha_{amp} = (4^o$ and $8^o)$ with $k = 0.025$.

2 Experimental Setup

Experiments have been completed in a closed circuit wind tunnel with a 152mm × 152mm cross-sectional area at the test section. A 55 mm chord length, c, S822 airfoil was actuated using a brushless AC servo motor and controller. Smoke wire flow visualization was used for capturing flow field images with high speed images obtained with a Photron SA1.1 high speed camera at 250 fps and 5000 fps and still images were obtained with a Nikon D70 digital SLR camera. Due to rapid smoke evaporation images were obtained at a low flow speed (Re=36,000 and 86,000). For the static cases, images were taken with two different photography setups. Laser Doppler Anemometry (LDA) was used for the wake study and mean drag measurements. A Dantec FiberFlow LDA system including a 310 mm focal length lens with a 5 W Argon laser was used for the measurements at the mid-span of the airfoil at $x/c = 1.25$ from the trailing edge.

A pressure gradient exists in the wake of the airfoil which can be replaced by the Navier-Stokes relation [8]. Moreover, fluctuations in the wake are important and can be measured by Root Mean Square (RMS) method (u_{RMS} and v_{RMS}). Mean velocity (u_{mean}) in the wake of the airfoil can be averaged (u_{avg}) based on all phases for a dynamic cycle, where the range of the phases are $[-4^o\ 4^o]$ or $[-8^o\ 8^o]$ according to the amplitude of the oscillations. By considering the pressure term and fluctuations while $U(h) \neq U_\infty$ ($[-h \quad h]$ is the measurement domain in the y coordinate) [8], the mean drag coefficient was calculated using the following equation:

$$C_{Dmean} = \frac{2}{c}\int_{-h}^{+h}\left[\frac{u_{avg}(y)}{U_\infty}(1-\frac{u_{avg}(y)}{U_\infty}) + \frac{1}{2}(1-\frac{U(h)}{U_\infty})(1-\frac{u_{avg}(y)}{U_\infty}) \right.$$
$$\left. -(\frac{u_{RMS}(y)}{U_\infty})^2 + (\frac{v_{RMS}(y)}{U_\infty})^2 - \frac{1}{2}(1-\frac{U(h)^2}{U_\infty{}^2})\right] \mathrm{dy}. \qquad (1)$$

3 Numerical Setup

An oscillating S822 airfoil was simulated using ANSYS Fluent 12. A C grid layout with 2×10^5 cells was used while $Y^+ \leq 1$. A dynamic mesh was employed to oscillate the entire mesh. A transition SST model, the $SSTk-\omega$ model and correlation-based transition model of Menter et al. [9], simulated the laminar-turbulent transition. All simulations were run over 16 CPUs in parallel.

4 Results and Discussions

4.1 Flow Visualization

As a preliminary work, numerical and experimental flow visualization are presented for static and dynamic cases when $Re < 10^5$. Fig. 1a,b show the Karman vortex-shedding patterns for $Re = 36,000$. The behavior of the vortex-shedding of the wake and the airfoil boundary layer are strongly coupled. For higher angles of attack, since the separation points move forward [10], the trailing edge boundary layer thickness increases resulting in larger shedding vortices with lower shedding frequencies. At the same Re for the dynamic case with $\alpha = 8\sin(\omega t)$ and $k = 0.025$, there is no evidence of the shedding vortex in Fig. 2. This indicates the flow structure of the wake for the oscillating and static airfoils are completely different. The shedding vortices for the static case disappear with increasing Re. Fig. 1c shows that at $Re = 86,000$, vortices in the wake are not clear and the simulation does not predict vortex shedding. Clear photographs for higher velocities with this technique are difficult and then based on Fig. 1c and simulations, not shown here, it is concluded that the wake of the static and dynamic cases for $Re = 10^5$ will be similar in terms of no shedding frequencies.

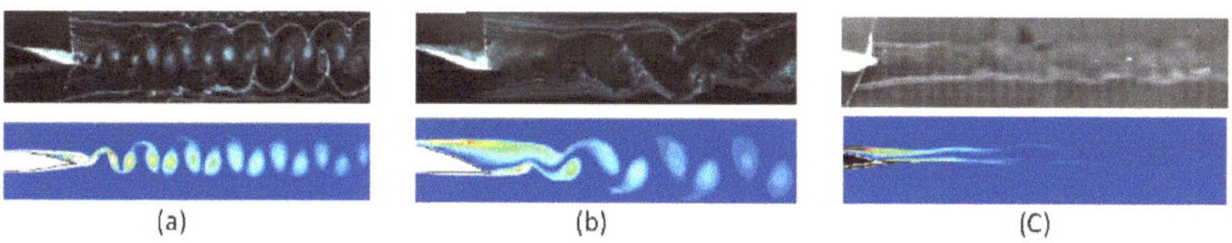

Fig. 1 Static flow patterns; upper: smoke wire photographs [11]; lower: numerical simulation; a) $\alpha = 0^o$, $Re = 36,000$; b) $\alpha = 8^o$, $Re = 36,000$; (c) $\alpha = 0^o$, $Re = 86,000$

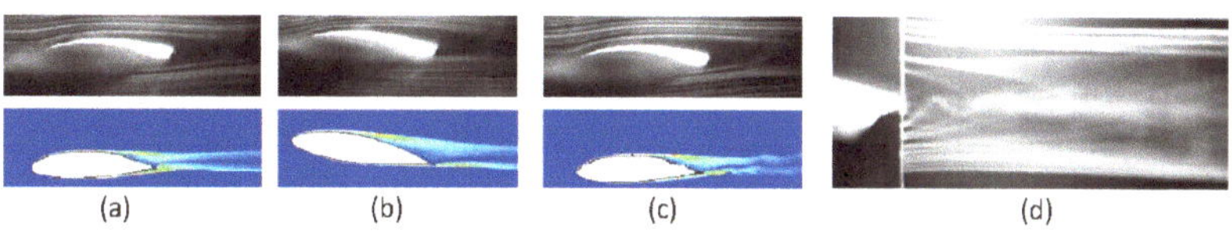

Fig. 2 Dynamic flow patterns for $Re = 36,000$, $\alpha = 8\sin(\omega t)$ and $k = 0.025$; a) $\alpha = 0^o \uparrow$ (upstroke); b) $\alpha = 8^o$; (c) $\alpha = 0^o \downarrow$(downstroke); (d) smoke wire located in the trailing edge for $\alpha = 0^o \uparrow$

4.2 Wake Study

Fig. 3 illustrates the averaged mean velocity for dynamic cases with amplitude 4° and 8° from LDA measurements and numerical simulation. The agreement between

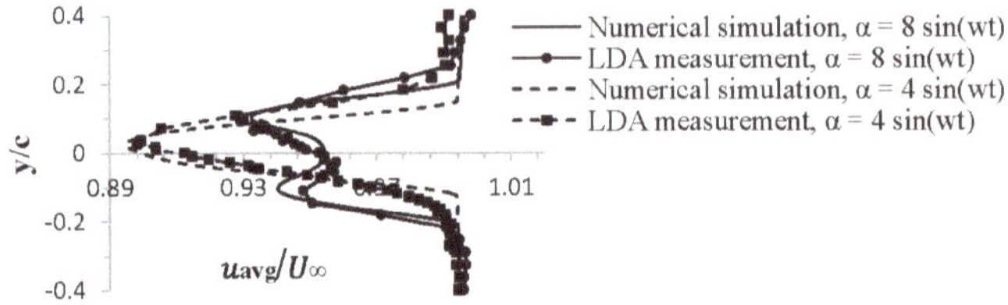

Fig. 3 Averaged streamwise mean velocity profiles ($x/c = 1.25$ and $Re = 10^5$)

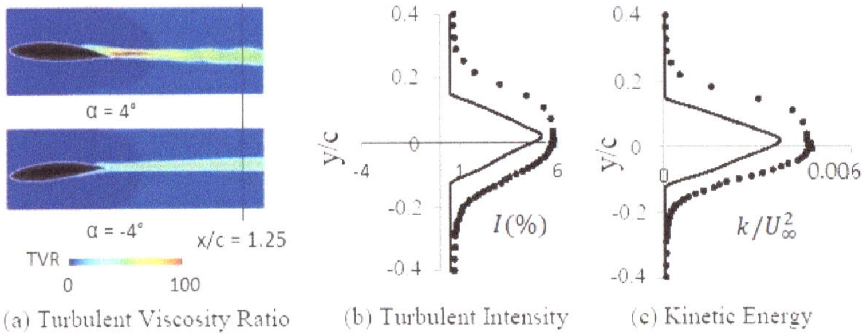

Fig. 4 Turbulent parameters for $\alpha = 4\sin(\omega t)$ ($Re = 10^5$); (a) Turbulent Viscosity Ratio (b) averaged turbulent intensity ($x/c = 1.25$) (c) non-dimensional averaged kinetic energy ($x/c = 1.25$); solid line: numerical results; dotted line: LDA measurement

the numerical and experimental results is obvious in this figure. Increasing the amplitude changes the structure of the averaged mean velocity from one peak structure to a double peak structure with a lower velocity deficit. The results show that a high amplitude causes a broader velocity wake profile with a smaller velocity deficit at most phase angles, not shown here.

Fig. 4a may indicate that for dynamic cases at high angle of attack laminar to turbulent transition in the wake occurs when the turbulent viscosity ratio $(TVR) \geq 100$. Fig. 4b, c show the averaged turbulent intensity and kinetic energy profiles at $x/c = 1.25$. The discrepancy between numerical and experimental results show that the Transition SST method predicts a narrower and less turbulent wake compared to the experimental results. Numerical turbulent intensity profiles for each phase and experimental turbulent intensity profiles for the static case are shown in Fig. 5. For all phases, a double peak structure is obvious. Although the numerical method underpredicts the turbulent variables, the comparison between dynamic and static cases reveals a narrower, but more turbulent wake for dynamic cases.

4.3 Aerodynamic Loads

Fig. 6 shows the range of the unsteady loads that the blade element experiences at each cycle. For the selected amplitudes and reduced frequency reported here the aerodynamic loads are very close to the static ones, but the unsteady behavior of

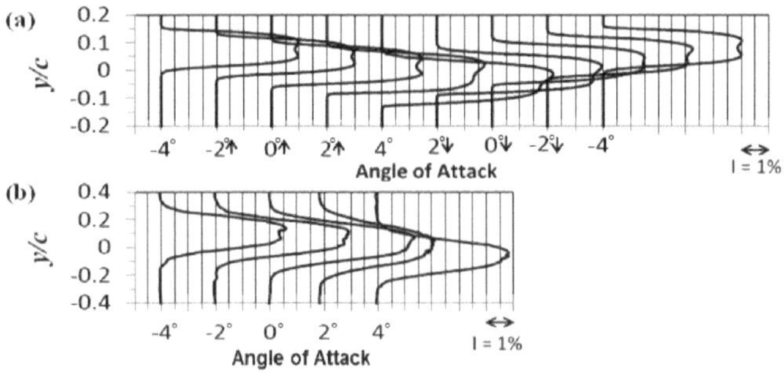

Fig. 5 Turbulent intensity profiles ($x/c = 1.25$ and $Re = 10^5$) (a) $\alpha = 4\sin(\omega t)$, numerical simulation (b) static case, LDA measurement

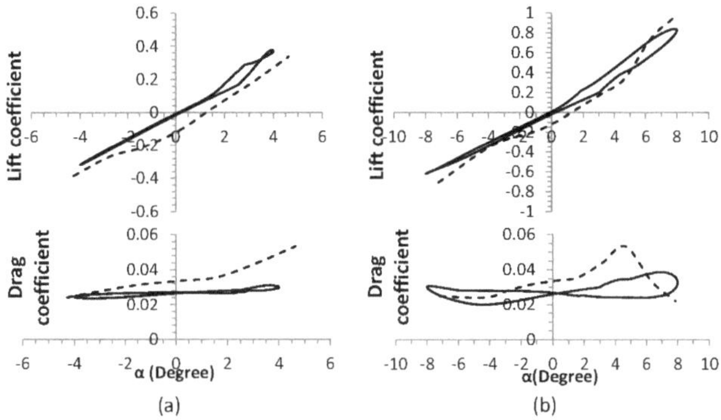

Fig. 6 Aerodynamic loads ($Re = 10^5$); Solid lines: dynamic cases from numerical simulation; Broken lines: static cases from experimental result [12]; (a) $\alpha = 4\sin(\omega t)$, (b) $\alpha = 8\sin(\omega t)$

the loading would imbalance the rotor. At each cycle, the mean drag value from the LDA measurement is also determined from equation 1 where $x/c = 1.25$. For $\alpha = 4\sin(\omega t)$, the mean drag values from numerical and LDA measurement are 0.0268 and 0.0275, respectively. For $\alpha = 8\sin(\omega t)$, the mean drag values from numerical and LDA measurement are also 0.0282 and 0.0323.

5 Conclusions

This study has revealed that boundary layer characteristics of the wakes for dynamic and static cases are completely different at low Reynolds numbers but at $Re = 10^5$, they are similar in terms of no shedding vortices. Increasing the amplitude of the oscillations makes a thicker boundary layer resulting in a low velocity deficit. Although there is a good agreement between numerical and experimental results, the transition SST model slightly underpredicts the turbulent parameters but it can predict the laminar-turbulent transition of the wake at the low amplitude and the low reduced frequency. Dynamic loads are close to the static values as expected from attached flows, but present the unsteady aerodynamic coefficients that the blade will face at each cycle.

Acknowledgements. The authors would like to acknowledge the support of the Natural Sciences and Engineering Research Council of Canada (NSERC), the facilities of the Shared Hierarchical Academic Research Computing Network (SHARCNET:www.sharcnet.ca) and Compute/Calcul Canada which made this work possible, Stephen Orlando for the static smoke wire photographs and Pauline Lambert from INSA de Lyon, France, for her assistance in taking LDA measurements.

References

1. Hansen, A.C.: Yaw Dynamics of Horizontal Axis Wind Turbines. NREL/TP-442-4822. National Renewable Energy Laboratory, Golden (1992)
2. Leishman, J.G.: Challenges in modelling the unsteady aerodynamics of wind turbines. Wind Energy 5, 85–132 (2002)
3. Narayana, M., Putrus, G.A., Leung, P.S., Jovanovic, M., McDonald, S.: Development of a model to investigate the yaw behaviour of small horizontal axis wind turbines. In: Proc. IMechE, vol. 226, pp. 86–97 (2012)
4. Maeda, T., Kamada, Y., Suzuki, J., Fujioka, H.: Rotor blade sectional performance under yawed inflow conditions. J. Sol. Energy Eng. 130, 031018-1-7 (2008), doi:10.1115/1.2931514
5. Wright, A.K., Wood, D.H.: Yaw rate, rotor speed and gyroscopic loads on a small horizontal axis wind turbine. Wind Eng. 31, 197–209 (2007)
6. Haans, W., Sant, T., van Kuik, G., van Bussel, G.J.W.: Measurement of Tip Vortex Paths in the Wake of a HAWT under Yawed Flow Conditions. J. Sol. Energy Eng. 127, 437–596 (2005)
7. Somers, D.M.: The S822 and S823 Airfoils, NREL Subcontractor Report, 500-36342 (2005)
8. Bohl, D.G., Koochesfahani, M.M.: MTV measurements of the vortical field in the wake of an airfoil oscillating at high reduced frequency. J. Fluid Mech. 620, 63–88 (2009)
9. Menter, F.R., Langtry, R., Volker, S.: Transition modeling for general purpose CFD codes. Flow Turbul. Combust. 77, 277–303 (2006)
10. Jung, Y.W., Park, S.O.: Vortex-shedding characteristics in the wake of an oscillating airfoil at low Reynolds number. J. Fluids Struct. 20, 451–464 (2005)
11. Orlando, S.M.: Laser Doppler anemometry and acoustic measurements of an S822 airfoil at low Reynolds numbers. Master's thesis, University of Waterloo, Waterloo, Canada (2011)
12. Selig, M.S., McGranahan, B.D.: Tests of six airfoils for use on small wind turbines, NREL/SR-500-34515, National Renewable Energy Laboratory (2004)

Comparative Measurements of the Effect of a Winglet on a Wind Turbine

Drew Gertz, David A. Johnson, and Nigel Swytink-Binnema

Abstract. A 3.3 m diameter variable speed wind turbine and rotor has been designed, fabricated and tested with exchangeable blade tip capability. This rotor was custom designed utilizing the NREL S83X series airfoils for the University of Waterloo Wind Energy (UWWE) research facility with blades allowing the outer 10% of the blade to be exchanged. The winglet was found to have a bell-shaped power augmentation profile, with a broad peak between 6.5 m/s and 9.5 m/s where power was increased by 5% to 8%. These power augmentation figures matched closely with the findings in the literature that suggested increases of 2% to 8% are possible.

1 Introduction

The designs of current horizontal axis wind turbine (HAWT) blade tips vary depending on scale but are for the most part simple shapes within the blade plane.

Simple alterations to the standard tip have been found to produce desirable bending load alleviation[1]; out of plane shapes, such as the endplate and winglet, can possibly augment the power output. Studies on small-scale rotors have found that the power produced by a wind turbine can be increased by 4 to 9% using such devices [2, 3]. Computational studies have also been performed on large rotors (MW-scale), and predict that power can be augmented by 2 - 3% at such a scale [4, 5, 6, 7, 8, 9]. Three-dimensional effects are significant on real airfoils and rotating airfoils such as wind turbine blades where the production of a tip vortex reduces the lift and induces drag. The majority of effort on wing tip losses have been directed at aircraft wings, and altering the geometry of the tip changes the effect of tip vortices. A geometric description of a winglet is shown in Figure 1 and is defined by the cant angle, height, twist, sweep, toe, and is also influenced by the chord distribution or

Drew Gertz · David A. Johnson · Nigel Swytink-Binnema
Wind Energy Group, Department of Mechanical and Mechatronics Engineering, University of Waterloo, Ontario, Canada
e-mail: {dpgertz,da3johns,nswytink}@uwaterloo.ca

M. Hölling et al. (eds.), *Wind Energy – Impact of Turbulence*, Research Topics in Wind Energy 2,
DOI: 10.1007/978-3-642-54696-9_18,

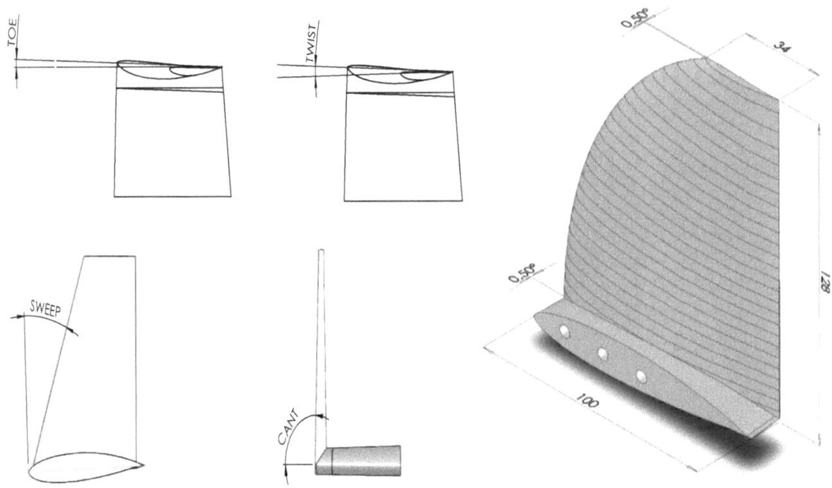

Fig. 1 (left) Geometric definition of winglet (right) Winglet Geometry (Dimensions in mm)

planform (rectangular, tapered, elliptical, etc.), airfoil, and location (pressure or suction side of wing).

Prior wind turbine winglet studies

The majority of published studies on wind turbine winglets have been numerical in nature. The consensus seems to be that it is possible to augment the power production of a turbine using winglets by means of a reduction in induced drag and hence greater torque. Table 1 summarizes the relevant studies and their results. The winglet designs outlined differ greatly, but there are some general trends. The power augmentation due to the winglet, P_{aug}, defined as the percentage change in the maximum power coefficient, has been predicted to be as high as 8% and as low as 2%, depending on the scale of the turbine. Documented experimental studies on wind turbine winglets in the open literature are scarce. Gyatt and Lissaman [10] report on three wing tip modifications that indicated that no increase in performance. A tip vane study on small turbines [3] documented tipvanes P_{aug} positive up to 9% at the optimal tip speed ratio and negative elsewhere.

2 Winglet Design

The winglet design shown in Figure 1 was based on literature discussed above. The winglet location was placed on the suction side, while the cant angle was chosen at 90°. The selection of height, root chord, and tip chord are important to manage the conflicting requirements of wetted-area penalty and Reynolds number ($Re = \frac{\rho U_\infty c}{\mu}$ where ρ and μ are the air density and dynamic viscosity, U_∞ is the average freestream velocity and c is the airfoil chord). Drag coefficients increase with

Table 1 Overview of numerical studies on wind turbine winglets

Source, Year	Cant (°)	Height (%R)	Twist (°)	Sweep (°)	Toe (°)	Airfoil	P_{aug} (%)	T_{aug} (%)	C_{Pmax}
[11] 1998	80	10	0	0	0	NACA 0012	8	8	NA
[8] 2006	90	1.5	-2	0	NA	NA	1.7	1.8	NA
[4] 2007	90	3-4	13	NA	5	Riso B118	2.9	3.1	.53
[9]2007	90	4	4	0	NA	NA	2.77	3.55	NA
[5] 2007	90	2	NA	0	NA	Riso B-15	2.8	3.5	.53
[6] 2008	90	5.4	NA	0	NA	NA	2.6	NA	0.5
[7] 2008	90	1-4	0-8	30	0	Riso B-15	2.2 (h-2%)	4.0 (h-2%)	NA
[2] 2009	90	10	8	NA	1	S809	3.5	NA	0.35

decreasing *Re*, and *Re* is proportional to chord length and greater chord lengths are better for *Re* considerations. Thus an optimal design will maintain greater chord lengths to satisfy *Re* concerns while also keeping the wetted area within reason. A winglet height of 8% was implemented and an elliptical planform was chosen. The root chord ratio, which is the ratio of the winglet root chord to blade tip chord, was set to 1 to maximize *Re* along the winglet span as well as ease of attachment. The winglet airfoil, the PSU 94-097[12] operated sufficiently at *Re* as low as 6×10^4 that then set the tip chord length to 34 mm and thus the taper ratio to 0.34. The toe angle provides the initial angle of attack and twist will control this angle as a function of winglet height. In a wind turbine application, it may be difficult to find an optimum toe angle that will operate at all flow conditions and the chosen airfoil has an impact on the decision. For example, the PSU 94-097 airfoil utilized here has a -5° zero lift angle of attack [12]. A toe of -0.5° and twist of -0.5° were chosen for this design.

3 Experimental

The experiment was performed at the University of Waterloo Wind Energy research facility (UWWE), an open-looped tunnel. Conditioned flow enters a 15.4 m wide, 19.5 m long test area that is 7.8 m high at the sides and 13 m high at the peak exiting the test area through a 7.9 m $\times$ 7.9 m door open to atmospheric air [13, 14]. A sonic 3D anemometer was used to measure the wind velocity and turbulence levels during testing and was located 1.5 m downstream from the exit plenum. A velocity correlation between that location and the rotor plane, which is 8 m downstream from the exit plenum, was used to determine the velocity at the rotor plane[15]. An uncertainty analysis revealed that the representative uncertainty in the velocity at the rotor plane was no more than $\pm 3.8\%$. Based on an analysis of data collected the freestream turbulence intensity, $I_T = (\sigma_U / U_\infty)$, was found to vary from 5.9% to 6.2% depending on the wind speed.

The planar blade length of 1.6 m was specifically developed using the NREL S83X series airfoils[16] for the UWWE research facility with a tip speed ratio λ of 5. The blade fabrication was split into a main blade and tip section in order to evaluate different blade tip shapes[13]. The surface oil technique was utilized to assist in understanding the winglet performance and visualize the surface flow. An electrical system was utilized including a load bank where voltage and current were measured that has been described in previous work[13, 14].

Experiments with 21 wind velocities, between 3.6 m/s and 11.0 m/s (nominal), were used to determine C_P vs λ. A common facility wind speed was used with most data obtained at constant rotor speed (200 rpm). The range of λ was extended by changing the rotor speed between 100 and 240 rpm. Sixty independent trials were obtained of 3 minutes duration at each operational point (0.3 Hz) and subsequently averaged.

4 Results and Discussion

Based on the power generation measurements and visualization outlined previously, the performance of the winglet in comparison to the standard tip blades could be evaluated.

Surface Oil Flow Visualization

The images taken of the outward-facing and hub-facing winglet surfaces at 8 m/s wind are shown in Figure 2. Region A on the outward-facing surfaces (pressure side) is a laminar region which then transitions to turbulent throughout region C. Region D is only different from C in that this region marks the flow which diverges off the tip of the winglet. Region B is a separated vortex region due to the flow leaving the pressure side of the planar blade. On the hub facing (suction) surface Region E is a laminar boundary layer which separates just beyond the dark region of low shear (region F) indicated by the oil accumulation line and transition to turbulence and reattachment occur again very soon. Over the hub-facing surface of the winglet, the pattern is much less clear although there is no definite separation region. The final region is Region G; it coincides with the area of the corner vortex also seen on the suction side of the planar portion of the blades.

Rotor performance with winglets

In addition to the rotor with standard tips, that is, the baseline rotor, performance data was gathered with the winglets for comparison. The data was gathered and processed using the same methods as for the standard tip and Figure 3 shows the percentage power augmentation over the operating range. Considering the literature indicated power augmentation of 2% to 8% could be expected, the experimental results having peak values between 5% and 8% are consistent. The shape of the power augmentation curve is also consistent with the related studies [3].

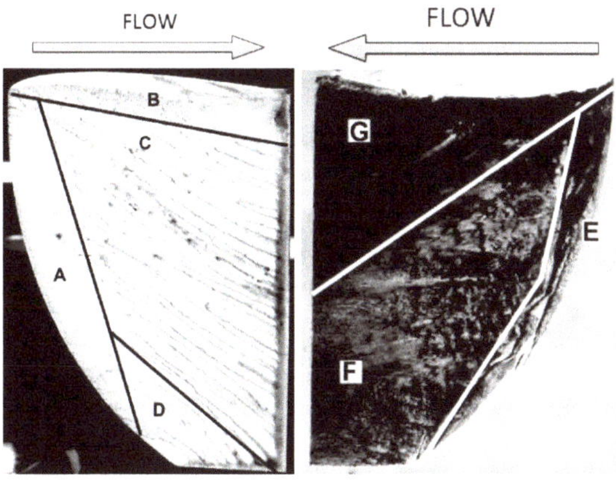

Fig. 2 Winglet surface oil visualization (**Left**) Outward-facing surface. Regions were interpreted as follows: (A) laminar region with transition; (B) planar tip circulation/vortex; (C) attached turbulent flow; and (D) flow diversion (**Right**) Hub-facing surface. (E) laminar; (F) attached turbulent flow; and (G) corner vortex.

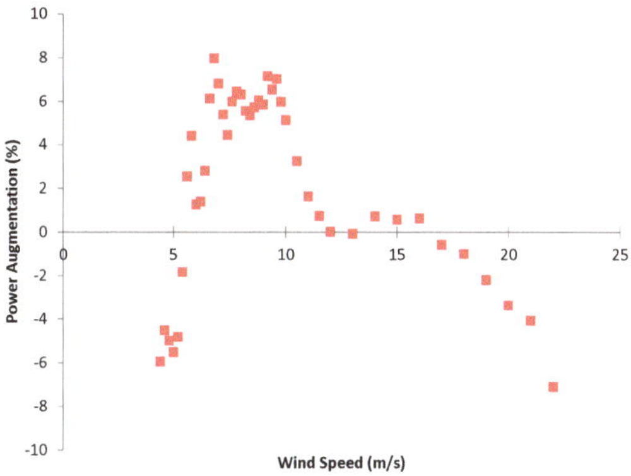

Fig. 3 Winglet power augmentation over planar blade versus wind speed

5 Conclusions

A 3.3 m diameter variable speed wind turbine and rotor has been designed and tested with exchangeable blade tip capability. This rotor was custom designed for the UWWE research facility with blades that allow the outer 10% of the blade to be exchanged. This allowed for the experimental characterization of the performance of the standard rotor and winglets.

Power production tests were completed covering a range of mean wind speeds, 3.6 m/s to 11 m/s nominal, and turbine rotor shaft rotational rates, 100 rpm to 240 rpm. The effect of the winglet was quantified by repeating the test that was executed with the standard tip for each winglet. The winglet was found to have a bell-shaped power augmentation profile, with a broad peak between 6.5 m/s and 9.5 m/s where power was increased by 5% to 8%. These power augmentation figures matched closely with the findings in the literature.

Acknowledgements. The authors would like to thank the Natural Sciences and Engineering Research Council (NSERC) of Canada for their financial support.

References

1. Antoniou, I., Madsen, H., Paulsen, U.: A theoretical and experimental investigation of new tip shapes. In: Proc. Euro. Comm. Wind Energy Conf., Germany, March 8-12 (1993)
2. Chattot, J.: Effects of blade tip modifications on wind turbine performance using vortex model. Computers and Fluids 38, 1405–1410 (2009)
3. Shimizu, Y., Ismali, E., Katmada, Y., Maeda, T.: Power Augmentation of a HAWT by Mie-type Tip Vanes, considering Wind Tunnel Flow Visualization, Blade-AR and Re. Wind Engineering 27, 183–194 (2003)
4. Døssing, M.: Vortex lattice modeling of winglet on wind turbine blades. Tech. Rep. Risø-R-1621(EN), Risø National Laboratory (2007)
5. Gaunna, M., Johansen, J.: Determination of the maximum aerodynamic efficiency of wind turbine rotors with winglets. Journal of Physics: Conference Series 75 (2007)
6. Hjort, S., Laursen, J., Enevoldsen, P.: Aerodynamic winglet optimization. Tech. rep., Siemens Wind Power A/S (2008)
7. Johansen, J., Gaunna, M., Sørensen, N.: Increased aerodynamic efficiency on wind turbine rotors using winglets. No. 2008-6728. In: 26th AIAA Appl. Aero Conf. (2008)
8. Johansen, J., Sørensen, N.N.: Aerodynamic investigation of Winglets on Wind Turbine Blades using CFD. Tech. Rep. Risø-R-1543(EN), Risø National Laboratory (2006)
9. Johansen, J., Sørensen, N.N.: Numerical Analysis of Winglets on Wind Turbine Blades using CFD. In: European Wind Energy Conference and Exhibition, Milan, IT (2007)
10. Gyatt, G., Lissaman, P.: Development and testing of tip devices for horizontal axis wind turbines. Tech. Rep. NASA CR-174991, NASA (1985)
11. Imamura, H., Hasegawa, Y., Kikuyama, K.: Numerical analysis of the horizontal axis wind turbine with winglets. JSME Int. J. 41, 170–176 (1998)
12. Maughmer, M.D., Swan, T.S., Willits, S.M.: The design and testing of a winglet airfoil for low-speed aircraft. No. AIAA 2001-2478. In: 19th Appl. Aero. Conf., Anaheim, CA (2001)
13. Gertz, D., Johnson, D.: An evaluation testbed for wind turbine blade tip designs baseline case. International Journal of Energy Research 35, 1360–1370 (2011)
14. Gaunt, B.: Power generation and blade flow measurements of a full scale wind turbine. Master's thesis, University of Waterloo, Waterloo, ON (2009)
15. Gertz, D.: An evaluation testbed for alternative wind turbine blade tip designs. Master's thesis, University of Waterloo (2011)
16. Tangler, J., Somers, M.: NREL Airfoil Families for HAWTs, NREL/TP-442-7109. Tech. rep. (1995)

Lidar Simulations to Study Measurements of Turbulence in Different Atmospheric Conditions

Jörge Schneemann, Stephan Voss, Gerald Steinfeld, Davide Trabucchi, Juan José Trujillo, Björn Witha, and Martin Kühn

Abstract. Modern lidar technology promises a fundamental enhancement of wind velocity measurements for site assessment. Previous studies have shown good agreements between lidars and mast mounted sonic anemometers concerning measurements of the 10 minute average horizontal wind velocity in flat terrain but have shown substantial differences concerning the measurement of turbulence intensity. One of the main reasons for poor turbulence measurements is assumed to lie in the scanning technique called VAD, applied by lidars. In contrast to a sonic anemometer, a VAD-scanning lidar senses the wind field at different positions along a circle. This is centred at the target point, and with a radius three orders of magnitude larger than the typical size of an anemometer. The resulting temporal and spatial averaging by the VAD scan influences the turbulence measurements. To understand the effects of different VAD scanning configurations and of the atmospheric condition on measuring turbulence, a numerical lidar scanner simulator was used. The influence of the VAD cone angle and the stability of the used LES generated wind fields were studied. The results show a high dependency on the used cone angle and the atmospheric stability.

1 Introduction

Wind energy is seen as one of the most important sources in the sustainable energy mix of the future. Although site assessment of new locations suitable for the

Jörge Schneemann · Gerald Steinfeld · Davide Trabucchi · Juan José Trujillo · Björn Witha · Martin Kühn
ForWind - Universität Oldenburg, Ammerländer Heerstrasse 136, Oldenburg, Germany
e-mail: j.schneemann@uni-oldenburg.de

Stephan Voss
TU Berlin, Strasse des 17. Juni, Berlin, Germany
e-mail: sgt.pepper1975@mailbox.tu-berlin.de

M. Hölling et al. (eds.), *Wind Energy – Impact of Turbulence*,
Research Topics in Wind Energy 2,
DOI: 10.1007/978-3-642-54696-9_19,

installation of wind turbines is an established process, it is still very costly. For site assessment according to IEC standards, measurements of the vertical wind profile up to hub height are required. Additionally, turbulence profiles are calculated from the wind data. The common way to measure these wind profiles is the use of met masts, which get more expensive with increasing hub height. Modern wind profilers using the remote sensing technique lidar (Light Detection And Ranging) offer an easy alternative for the measurement of vertical wind profiles. Several studies show good agreements of the mean wind velocity and direction, comparing several commercial lidars with mast mounted sonic anemometers in flat terrain (e.g. [2, 3]). In contrast, the measurement of turbulence with lidars was found to compare badly in recent studies [1, 4]. Comparisons between a mast mounted sonic anemometer offshore with a lidar placed next to it, showed significantly higher turbulence intensities. Spectral analysis revealed higher power of the lidar signal than the sonic anemometer signal in the frequency range between 0.004 Hz and 0.2 Hz [1]. Moreover, the systematic errors in the second order moments of wind speeds, measured by two different commercial lidars in different atmospheric conditions, were modelled and compared to measurements in another study [4]. The simulations, as well as the measurements, showed high errors in variance measurements derived from the lidar data compared to the sonic anemometer. The results showed a high dependency on the atmospheric stability. This motivates the present work, where it is tried to optimize the scanning procedure applied by a lidar, and to understand the influence of atmospheric stability on the lidar measurement, by means of numerical simulations. The influence of VAD (Velocity Azimuth Display, the standard scan strategy for wind lidars) configurations and of the atmospheric stability on the measurement of the vertical wind profile, with a focus on turbulence, is studied. For this purpose a comprehensive lidar scanner simulator (LiXim) and different wind fields, generated by large eddy simulations (LES), with different atmospheric conditions, were used to simulate lidar measurements with variable configurations. The VAD configuration used is the standard four beam set up, employed e.g. in the wind profiler Windcube®, developed by the French company Leosphere. This VAD scan derives the wind vector by measuring the line of sight (LoS) wind velocity in four different perpendicular directions along the surface of a cone. The effect of higher vertical wind turbulence is evaluated by using LES generated wind fields with different atmospheric conditions (neutral, unstable and very unstable) for the lidar simulations. The consecutive analysis is done by calculating the ratio of variance from the lidar data to the variance of the corresponding wind field data and by spectral analysis.

2 Methods

Lidar simulator

The lidar scanner simulator LiXim, developed at Oldenburg University, is capable of realistic simulations of pulsed lidar's line of sight measurements in a given wind field, using a simplified physical model. The number of selectable range gates is just limited by the resolution of the input wind field. The spatial resolution in

LoS-direction is about 40 m. Volume sampling is modelled by a convolution of the range gate and the pulse shape function. In discreet wind fields it is interpolated between relevant points. Different lidar scan strategies can be used for the simulations. Details on the implementation and a verification of the procedure are found in [5].

Large eddy simulation wind fields
The three wind fields with different atmospheric conditions "measured" by the lidar simulator were generated by the LES code PALM (Table 1).

Table 1 Specifications of the wind fields used for the simulations

atmospheric condition	neutral	unstable	very unstable
dimension [m^3]	400 x 720 x 264	400 x 2000 x 300	400 x 2000 x 300
resolution [m^3]	4 x 4 x 4	4 x 4 x 4	4 x 4 x 4
temporal length [s]	760	600	600
temporal resolution [Hz]	2	2	2
near surface heat flux [K·m/s]	0	0.01	0.1

VAD Scan
The simulated lidar performs a conical VAD scan with four LoS measurements according to Figure 1. The middle of the measurement volume in LoS direction is taken as the measurement point in the following. The components of the wind vector

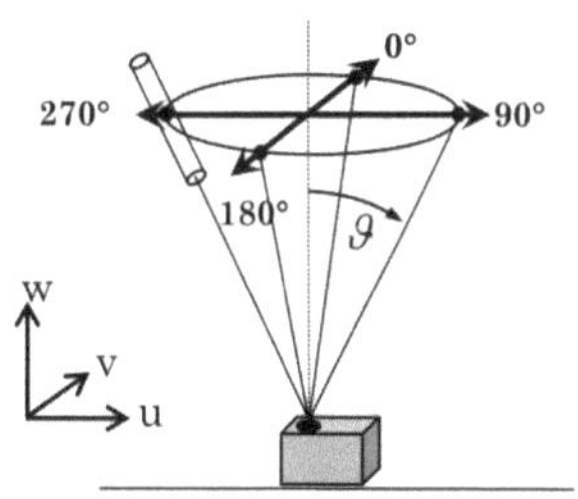

Fig. 1 Sketch of the used VAD scan: For the simulations we choose the VAD-scan used by the Windcube, a common lidar. It performs a conical scan with the cone angle ϑ. Measurements are done in the cardinal points.

$V = (u, v, w)$ are calculated from the LoS measurements V_{LoS} assuming a homogeneous wind field ($u = u_{90} = u_{270}$; $v = v_0 = v_{180}$ and $w = w_0 = w_{90} = w_{180} = w_{270}$), using

$$u = \frac{V_{LoS\ 90} - V_{LoS\ 270}}{2\sin\vartheta}, \qquad v = \frac{V_{LoS\ 0} - V_{LoS\ 180}}{2\sin\vartheta}$$
$$w = \frac{V_{LoS\ 0} + V_{LoS\ 90} + V_{LoS\ 180} + V_{LoS\ 270}}{4\cos\vartheta}, \qquad (1)$$

where ϑ is the cone angle. The theory of the VAD scan can be found e.g. in [6].

Simulation of lidar measurements
Since the temporal length of the used LES wind fields is limited to 600 s of wind data, several lidars were simulated simultaneously in different places of a wind field. Each lidar performed a VAD-scan with the same configuration, but the measurement points were rotated around a vertical axis to avoid a measurement overlap. Simulations of VAD-scans with different cone angles ϑ were performed.
Measurements are assumed instantaneous, the movement of the scanner to the next LoS is 1 s. To measure the full cone it takes 4 s. Values of the wind velocity are generated every second, using the newest set of LoS-measurements. Values for different altitudes (above the lidar) are calculated at once. Details of the different simulation setups can be found in Table 2.

Table 2 Setup of the simulated lidar measurements

cone angles [deg]	atmospheric conditions	heights [m]
5, 10 15, 24, 26, 28, 30, 32, 34, 40	neutral, unstable, very unstable	50, 100, 150, 200

Data analysis
To compare the turbulence measured by the lidar with the reference wind field, the second order moments (variances Var) of the wind vector components on intervals of length $\Delta t = 60$ s are calculated. Analogue to [4] the median (marker) and the first and third quartile (error bars) of Var(lidar)/Var(wind field) of the whole ensemble of Δt intervals is plotted (Figures 2 and 4). The deviation of this value from unity is the systematic error of the simulated lidar measurement.
As a second analysis, the power spectral density (PSD) of the lidar data and the corresponding wind field data is calculated for each $\Delta t = 60$ s interval and averaged afterwards (Figure 3).

3 Results and Discussion

Influence of the cone angle
Figure 2 shows the influence of the cone angle ϑ on the systematic error of the simulated lidar data. Results for cone angles below 15° are omitted, because of the large systematic errors. It can be clearly seen that the variance overestimation for u and v is decreasing for higher cone angles. The underestimation of the variance of w is almost constant. This behaviour is consistent within all atmospheric conditions and heights. Similar results, overestimations of u and v and a significant underestimation of w (height = 100 m, unstable atmosphere), were measured by [4].

Figure 3 shows the PSD of the horizontal wind $U_{hor} = \sqrt{u^2 + v^2}$ for the wind field with unstable stratification and of the lidar simulations for different cone angles ϑ.

The PSD of the LES wind field shows a good agreement with the power law over a wide range. The simulated lidar data show a significant power increase over all analysed frequencies for all cone angles. This overestimation of the PSD is decreasing with increasing cone angle of the VAD scan. The components of the wind vector

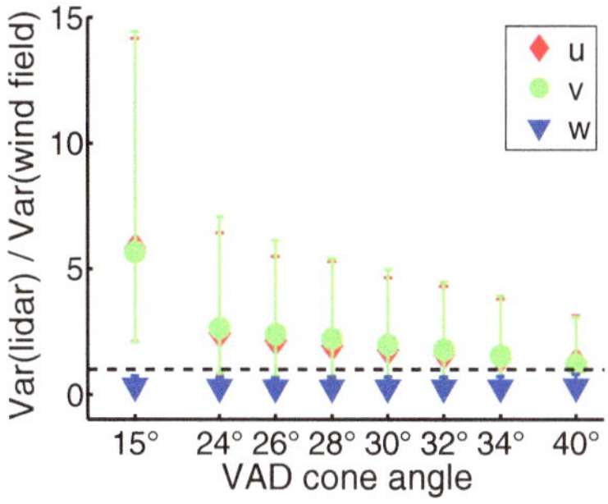

Fig. 2 Systematic errors of the simulated lidar measurements for different cone angles of the VAD scan (unstable atmosphere, height = 100 m). The horizontal dashed line marks unity, meaning no systematic error.

were analysed separately, as well. Consistent with the results shown in Figure 2, the overestimation of the PSD can just be seen for the horizontal components u and v, while the PSD is underestimated by the vertical component w for all frequencies in all conditions.

The minima in the spectral power density at 0.25 Hz and 0.5 Hz (as found in a free field experiment with a similar lidar by [2]) result from the time taken to measure a full scan (4 s) and a half scan (2 s). The lidar is "blind" for structures with these lengths.

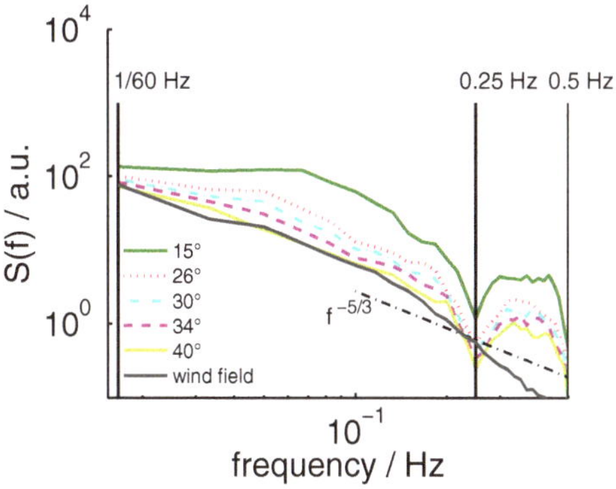

Fig. 3 Double logarithmic plot of the PSD of the horizontal component U_{hor} of the unstable wind field (height: 100 m) as well as the simulated lidars for different cone angles ϑ of the VAD cone. Additionally, a line with Kolmogorov's power law $f^{-5/3}$ is shown.

Influence of the atmospheric condition

The systematic errors of the lidar data for different atmospheric conditions is shown in Figure 4. Slightly increasing overestimations for u and v and a slightly increasing underestimation for w are found from neutral to unstable. These results hold valid for all simulated conditions. Similar results for u and v but no dependency of w on atmospheric stability were measured by [4].

Influence of the measurement height

The simulations indicate an increasing overestimation of u and v and an almost constant underestimation of w with increasing measurement height for all simulations (not shown). Similar dependencies except for the tendency of w to have increasing systematic errors with height were also measured by [4].

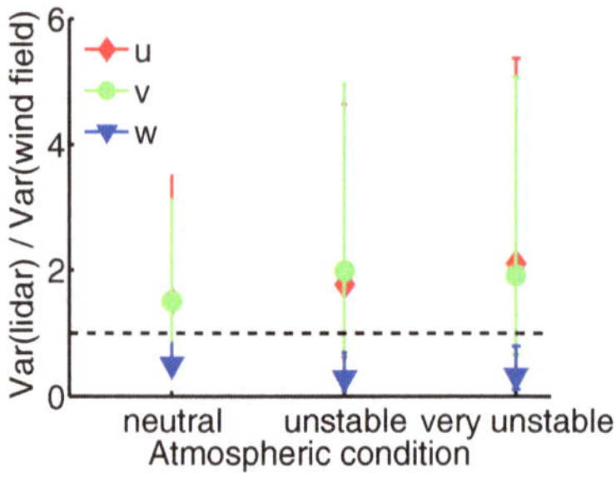

Fig. 4 The lidar's systematic error over the atmospheric stability for $\vartheta = 30°$ and 100 m height

4 Conclusions

Lidar's VAD scans with different cone angles were simulated in LES generated wind fields with different stabilities. The simulations revealed that the measurement of turbulence with pulsed lidar highly depends on the VAD scan's cone angle as well as on atmospheric stability.

Systematic errors of turbulence measurements on the horizontal wind components increase with decreasing cone angles and with decreasing atmospheric stability. The often used cone angle $\vartheta = 15°$ leads to very high systematic errors and seems not to be suitable for turbulence measurements. For $\vartheta = 40°$, the systematic error is already very low, so that the PSD curve fits the wind field's curve over a wide range. Except for deflections in the w component the simulated results agree with the measurements done by [4].

References

1. Cañadillas, B., Bégué, A., Neumann, T.: Comparison of turbulence spectra derived from LiDAR and sonic measuremnts at the offshore platform FINO1. In: Proceedings of the 10th German Wind Energy Conference, DEWEK 2010, Bremen, Germany (2010)
2. Cañadillas, A., Westerhellweg, A., Neumann, T.: Testing the Perfomance of a Ground-based Wind LiDAR System. DEWI Magazin. 38, 58–64 (2011)
3. Gottschall, J., Courtney, M.: Verification test for three WindCubeTM WLS7 LiDARs at the Høvsøre test site. Risø-R-1732(EN) Risø DTU, Roskilde, Denmark (2010)
4. Sathe, A., Mann, J., Gottschall, J., Courtney, M.: Can Wind Lidars Measure Turbulence? Journal of Atmospheric and Oceanic Technology 28(7), 853–868 (2011)
5. Trabucchi, D., Trujillo, J.J., Steinfeld, G., Schneemann, J., Machtaa, M., Cariou, J.P., Kühn, M.: Numerical assessment of performance of lidar WindScanners for wake measurements. Presented at the EWEA Annual event (2011)
6. Werner, C.: Doppler Wind Lidar. In: Weitkamp, C. (ed.) Lidar, vol. 102, pp. 325–354. Springer, Heidelberg (2005)
7. Westerhellweg, A., Cañadillas, B., Beeken, A., Neumann, T.: One Year of LiDAR Measurements at FINO1- Platform: Comparison and Verification to Met-Mast Data. In: Proceedings of the 10th German Wind Energy Conference, DEWEK 2010 (2010)

Part IV

Flow Modeling

Vortex Particle-Mesh Simulations of Atmospheric Turbulence Effects on Wind Turbine Blade Loading and Wake Dynamics

Stéphane Backaert, Philippe Chatelain, Grégoire Winckelmans, and Ivan De Visscher

Abstract. We develop a Vortex Particle-Mesh (VPM) method for the Large Eddy Simulation of wind turbine wakes and coarse scale aerodynamics.

Based on the vorticity formulation of the Navier-Stokes equations, the approach combines the advantages of two discretizations. The particles handle the advection of vorticity, and exploit its compact support. The mesh is used for the evaluation of the differential operators and inside efficient Fourier-based Poisson solvers.

The blades are modeled by means of lifting lines. This treatment is similar to a vortex lattice method, although it is here immersed in the mesh and allows the evolution of shed vorticity into a fully turbulent flow.

We apply this tool to the investigation of atmospheric turbulence effects on the blade aerodynamics and on the wake behavior. To this end, the inflow boundary is feeding our VPM calculations with a turbulent vorticity field that has been computed in a preliminary LES of atmospheric turbulence.

1 Introduction

The envisioned developments in wind energy over the coming decade involve larger sizes for the devices and power plants, and higher efficiencies, too. These trends will make power conversion processes and the device structures even more sensitive to flow unsteadiness and the subsequent fatigue processes. In those respects, understanding the interactions between the atmospheric turbulence and wind turbine aerodynamics will be crucial.

Wind turbine aerodynamics have been the focus of numerical investigations relying on several techniques. Free wake Vortex Lattice methods offer an affordable means of evaluating the dynamics of the vortex shedding and the near wake; they cannot however handle the dissipation of this sheet and its transition into a three-dimensional fully turbulent flow. Actuator line techniques have been proposed and

Stéphane Backaert · Philippe Chatelain · Grégoire Winckelmans · Ivan De Visscher
Institute of Mechanics, Materials and Civil Engineering, Université catholique de Louvain, Place du Levant, 2, B-1348 Louvain-la-Neuve, Belgium
e-mail: stephane.backaert@uclouvain.be

M. Hölling et al. (eds.), *Wind Energy – Impact of Turbulence*,
Research Topics in Wind Energy 2,
DOI: 10.1007/978-3-642-54696-9_20, © Springer-Verlag Berlin Heidelberg 2014

include a body force term that accounts for the blade loading into a Navier-Stokes solver [7]. These techniques have been going through intensive development over recent years and have been applied to cases with a turbulent inflow [8].

Vortex methods can be seen to combine the aspects of free wake Vortex Lattice methods and actuator line techniques. Based on a vorticity formulation, they exploit the compactness of that quantity and allow the shedding of fine vortical structures. At the same time, they allow these flow structures to evolve into a fully turbulent flow, which makes them akin to a high fidelity variant of a free wake method, and an interesting tool for the validation and development of design codes (BEM or free wake).

We present work on a state-of-the-art variant of vortex methods that combines both particles and a mesh, exploiting the advantages of both discretizations. We here introduce the handling of a turbulent inflow, in vorticity form. We apply our approach to a single wind turbine, and subject it to an incoming Large Eddy Simulation result.

2 Methodology

2.1 Vortex Particle-Mesh Method

We consider a three dimensional incompressible flow and the Navier-Stokes equations in their velocity ($\mathbf{u}$)-vorticity ($\omega = \nabla \times \mathbf{u}$) form

$$\frac{D\omega}{Dt} = (\omega \cdot \nabla)\,\mathbf{u} + \nu \nabla^2 \omega \tag{1}$$

$$\nabla \cdot \mathbf{u} = 0 \tag{2}$$

where $\frac{D}{Dt} = \frac{\partial}{\partial t} + \mathbf{u} \cdot \nabla$ denotes the Lagrangian derivative and ν is the kinematic viscosity. Vortex methods discretize the vorticity field with particles, characterized by a position $\mathbf{x}_p$, a volume V_p and a strength $\alpha_p = \int_{V_p} \omega d\mathbf{x}$. Particles are convected by the flow field and their strength is modified to account for vortex stretching and diffusion.

Using the definition of vorticity and the incompressibility constraint the velocity field is computed by solving the Poisson equation (or Biot-Savart law)

$$\nabla^2 \mathbf{u} = -\nabla \times \omega\,. \tag{3}$$

This equation is solved with the following boundary conditions: an inlet-outlet direction (along z) and unbounded in the transverse directions (along x and y).

The mesh allows for additional advances. Differential operators (such as those for stretching and diffusion) are evaluated on the mesh using fourth order finite differences and the Poisson equation for velocity (Eq. (3)) is handled on the grid in Fourier space. The results of these calculations, i.e. the right-hand side of Eq. (1) and the velocity field, are then interpolated back from the grid onto the particles. Finally, the grid-based Poisson solver is also used to reproject vorticity onto a solenoidal field, i.e. enforcing

$$\nabla \cdot \omega = 0\,, \tag{4}$$

every few time steps (of the order of 50 for the present simulations). We refer to [2, 3] for details on the parallel implementation and the Poisson solver.

The resulting method handles the advection of vortical structures with no dispersion error, its time stability is not constrained by the classical CFL condition, and enforces the same boundary conditions as a BEM or Vortex Lattice method.

The sources of vorticity are modeled through a lifting line model, in a fashion similar to a Vortex Lattice technique. If the blade lift distribution is known, the bound circulation $\Gamma(r,t)$ and the shed vortex sheet can be obtained from the Kutta-Joukowsky equation and the solenoidal character of vorticity. In our particle-mesh setting, this shed vortical structure is actually an incremental change to an already three-dimensional vorticity field [1].

2.2 Turbulent Inflow

The implementation of a turbulent inflow has to preserve two properties of our VPM method: its Lagrangian treatment and ability to handle unbounded directions (Fig. 1).

First, our Lagrangian method requires this field to be transformed into particles, which will evolve according to the Navier-Stokes equations once inside the domain. A buffer region feeds particles into the domain at the target mean velocity; the particles' strengths are interpolated "on the fly" from the turbulent field. The velocity of the particles lying in this region progresses from the mean velocity for the newest ones to the velocity induced by the Biot-Savart law for those ready to be added to the computational domain.

Secondly, some precautions in the unbounded directions have to be taken while this field is being fed. Indeed, the evolution of the incoming turbulence close to unbounded boundaries might cause the initial domain to grow. This would cancel the advantage of the compactness of the vorticity support. A clipping function is applied to the strengths of the entering particles in the unbounded directions. The combination of this smooth function with the periodic reprojection of the overall vorticity field onto a solenoidal field preserves the compact support of the incoming turbulence and properties such as its isotropy or homogeneity.

The turbulent field is introduced periodically into the domain. Figure 2 shows the vorticity field generated by the feeding of a 2D long cubic homogeneous isotropic turbulent (HIT) field into a 2Dx2Dx8D domain. This HIT was generated by a spectral LES of forced turbulence [4]. We observe that the entrainment of null vorticity patches towards the center line. This calls for an initial domain large enough to push this effect away from the physics of interest.

The evolution of the characteristics of the HIT field is presented in Fig. 3. As expected, the turbulence exhibits a decay as it travels downstream. The velocity variances remain close, indicating the minor effect of the clipping on isotropy.

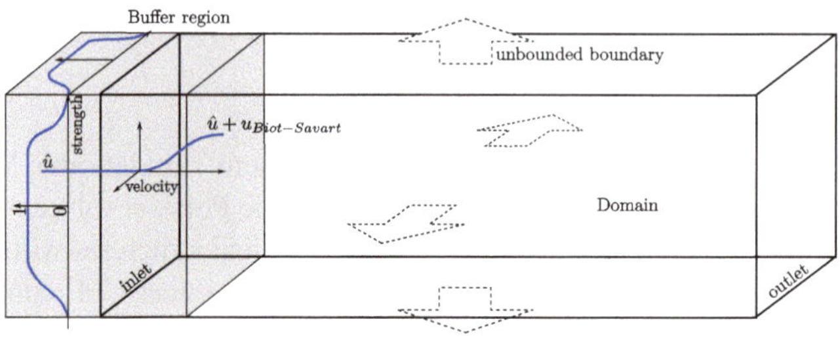

Fig. 1 Turbulent inflow and computational domain: vorticity clipping and velocity field transition in the buffer region

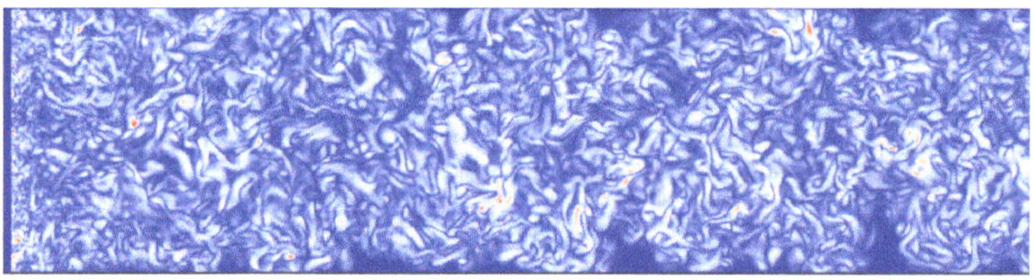

Fig. 2 Turbulent inflow: 2D cubic turbulent box fed into 2Dx2Dx8D domain, vorticity magnitude in the center plane

Fig. 3 Variances of velocity components. Two different subgrid scale (SGS) models were used: a second order HV model and the Smagorinsky model. The slight difference near the inlet is due to the mismatch between the SGS models used to compute the prescribed turbulent field and the one acting during the simulation.

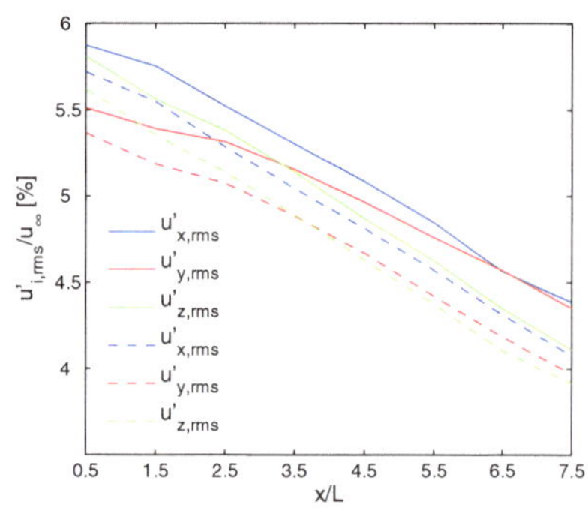

3 Results

We assess our methodology on the Tjaereborg wind turbine as this has been the benchmark in several works [5, 9] with a tip-speed ratio $\lambda = 7.07$.

The computational domain covers eight rotor diameters, with the wind turbine placed one diameter from the inlet boundary. The spatial resolution amounts to 64 points per blade, leading to a problem size of $256 \times 256 \times 1024$. This particular simulation covers 10 dimensionless times D/U_∞ and used 1024 cores for 24 hours.

The time-averaged circulation along the blades conserves the same distribution as in the case with a uniform inflow (Fig. 4(a)). However, its variance is not uniform.

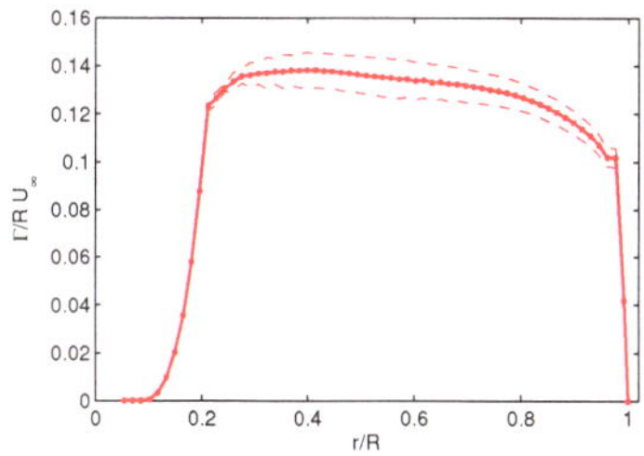

(a) Dimensionless circulation distribution: average (solid) and standard deviation (dashed)

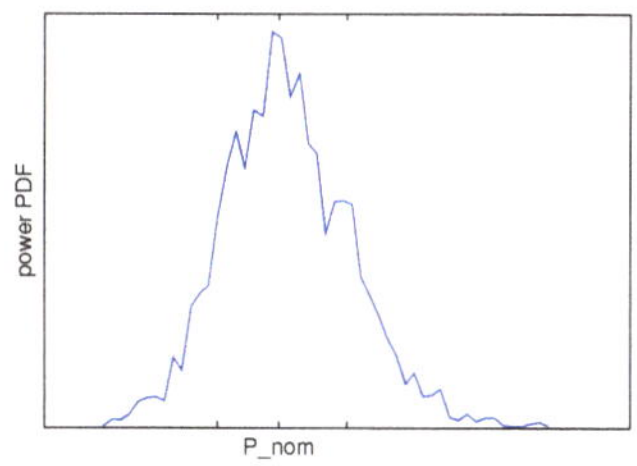

(b) Probability density function of power: the standard deviation σ is shown.

Fig. 4 Tjaereborg rotor: turbulence effects

The contributions of the variations of relative velocity and angle of attack vary along the blade: strong variation of the angle of attack near the root and higher influence of the rotation speed near the tip.

The integration of the forces acting on each blade element leads to the power probability density function (PDF) and to the PDFs of rotor fatigue contributions such as the azimuthal bending moment or the torque variation. The figure Fig. 4(b) shows indeed the effect of this turbulence on the produced power.

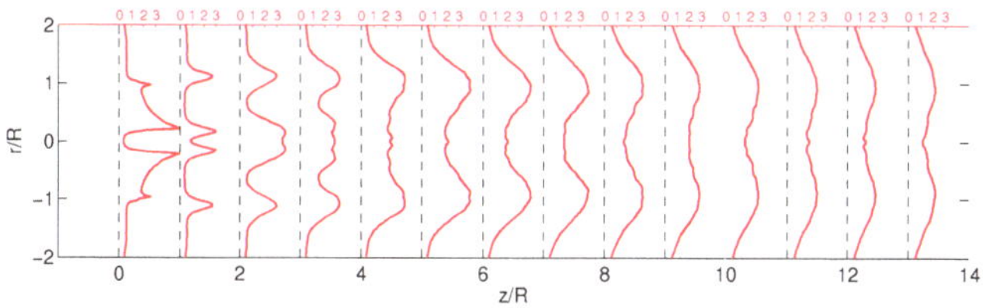

Fig. 5 Tjaereborg rotor: turbulent kinetic energy profiles $\langle u_x'^2 + u_y'^2 + u_z'^2 \rangle / 2U_\infty^2$ (%)

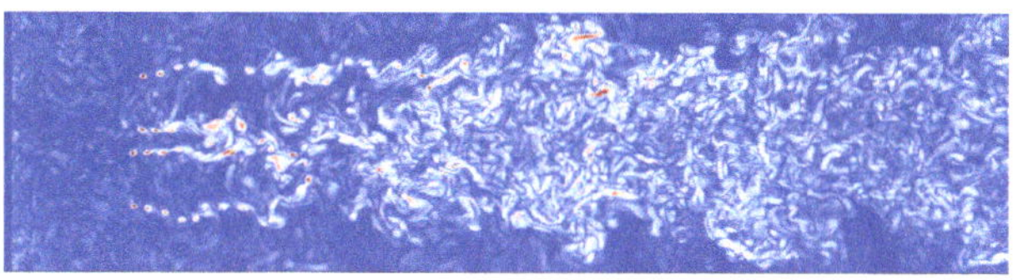

Fig. 6 Tjaereborg rotor: vorticity magnitude in center plane

The turbulent kinetic energy (TKE) profiles (Fig. 5) show that the slipstream is more turbulent in the near wake than in the case without turbulent inflow [1]. This is due to the more pronounced lateral oscillations of the tip vortices (Fig. 6). The instabilities of these structures have an impact on the length of the hub jet, which breaks down sooner.

4 Conclusions

We have developed a turbulent inflow technique for vortex particle-mesh methods. Our approach is Lagrangian and does not impose further constraints on the time step. Our results show the capability of the method to capture correctly the evolution of turbulent wakes as well as the turbulent effects on the rotor performance.

Areas of future work include multiple wind turbine investigations, specifically wake interactions and meandering, and the use of turbulent inflows more realistic than HIT such as synthetic turbulent flow fields [6].

References

1. Backaert, S., Chatelain, P., Bricteux, L., Winckelmans, G., Koumoutsakos, P.: Vortex particle-mesh methods with immersed lifting lines applied to the large eddy simulation of wind turbine wakes. Journal of Computational Physics (in preparation)
2. Chatelain, P., Curioni, A., Bergdorf, M., Rossinelli, D., Andreoni, W., Koumoutsakos, P.: Billion vortex particle direct numerical simulations of aircraft wakes. Computer Methods in Applied Mechanics and Engineering 197(13), 1296–1304 (2008)
3. Chatelain, P., Koumoutsakos, P.: A Fourier-based elliptic solver for vortical flows with periodic and unbounded directions. Journal of Computational Physics 229(7), 2425–2431 (2010)
4. De Visscher, I., Bricteux, L., Winckelmans, G.: Aircraft vortices in stably stratified and weakly turbulent atmospheres: simulation and modeling. AIAA Journal (2012) (accepted)
5. Ivanell, S., Sorensen, J.N., Mikkelsen, R., Henningson, D.: Analysis of numerically generated wake structures. Wind Energy 12(1), 63–80 (2009)
6. Mann, J.: The spatial structure of neutral atmospheric surface-layer turbulence. Journal of Fluid Mechanics 273, 141–168 (1994)
7. Sorensen, J.N., Shen, W.Z.: Numerical modeling of wind turbine wakes. Journal of Fluids Engineering - Transactions of the ASME 124(2), 393–399 (2002)
8. Troldborg, N., Sorensen, J., Mikkelsen, R.: Actuator line simulation of wake of wind turbine operating in turbulent inflow. Journal of Physics: Conference Series, 012063, 15 (2007)
9. Troldborg, N., Sorensen, J.N., Mikkelsen, R.: Numerical simulations of wake characteristics of a wind turbine in uniform inflow. Wind Energy 13(1), 86–99 (2010)

Wind Energy Applications of Unified and Dynamic Turbulence Models

Stefan Heinz and Harish Gopalan

Abstract. This study investigates linear and nonlinear dynamic LES and unified RANS-LES models obtained by stochastic analysis regarding the simulation of the atmospheric boundary layer. The advantages of dynamic LES methods are their accuracy, the ability to account for anisotropy and backscatter, and their computational stability. The advantage of unified RANS-LES models is their ability to provide results like LES for a fraction of the cost of LES. Two applications are considered: (i) turbulent boundary layer simulations on fine grids, which were performed to demonstrate the suitability of models for turbulent boundary-layer turbulence studies, and (ii) neutral boundary layer simulations on coarse grids, where subgrid models have to contribute significantly. On fine grids, all models perform comparably. On coarse grids, a nonlinear dynamic model provides the most accurate results. A nonlinear unified RANS-LES model is shown to be well applicable as a low cost alternative.

1 Introduction

There is a growing interest in using wind energy to provide electricity all around the globe. For example, a recent report by the Department of Energy suggests the possibility of providing 20% of the electricity in the U.S. by wind energy in 2030. A significant problem of existing wind farms is their relatively low efficieny. Due to the wind turbine wakes, the efficiency of wind farms is reduced by 20 to 50

Stefan Heinz
Department of Mathematics, University of Wyoming, 1000 E. University Avenue, Laramie, WY 82071, USA
e-mail: heinz@uwyo.edu

Harish Gopalan
Department of Mechanical Engineering, University of Wyoming, 1000 E. University Avenue, Laramie, WY 82071, USA
e-mail: hgopalan@uwyo.edu

M. Hölling et al. (eds.), *Wind Energy – Impact of Turbulence*, Research Topics in Wind Energy 2,
DOI: 10.1007/978-3-642-54696-9_21,

percent compared to turbines in isolation [1]. Therefore, there is a significant need for research related to the interaction of wind turbines with the surrounding air flow.

The only way to obtain a comprehensive understanding of wind farm processes is the use of numerical simulations. However, such simulations face significant problems related to the consistency of equations, the generality of equations, and the computational cost of simulations. A very promising approach to find general solutions for these questions is the development of turbulence models on the basis of stochastic analysis. Unified combinations of Reynolds-averaged Navier-Stokes (RANS) equations with large eddy simulation (LES) equations, and dynamic LES equations were presented recently by [2, 3]. The advantages of dynamic LES methods are their accuracy, the ability to account for anisotropy and backscatter, and their computational stability. The advantage of unified RANS-LES models is their ability to provide results like LES for a fraction of the cost of LES. Details about the characteristic features of these methods and applications to channel flow simulations can be found elsewhere [4, 5, 6].

The purpose of this paper is to report applications of these unified and dynamic methods to turbulent boundary-layer turbulence studies relevant to wind energy applications. The turbulence models applied will be described in Sect. 2. Section 3 deals with a presentation of the numerical methods applied and problems considered. The results obtained by means of unified and dynamic LES models are discussed in Sect. 4. Section 5 summarizes the conclusions of these studies.

2 Unified and Dynamic Model Equations

For the incompressible flow considered, the LES equations for filtered velocities $\widetilde{U}_i$, where $i = 1,2,3$, are given by the incompressibility constraint $\partial \widetilde{U}_i / \partial x_i = 0$ and the momentum equation

$$\frac{\widetilde{D}\widetilde{U}_i}{\widetilde{D}t} = -\frac{1}{\rho}\frac{\partial \widetilde{p}}{\partial x_i} + 2\nu \frac{\partial \widetilde{S}_{ik}}{\partial x_k} - \frac{\partial \tau_{ik}}{\partial x_k}. \tag{1}$$

Here, $\widetilde{D}/\widetilde{D}t = \partial/\partial t + \widetilde{U}_k \partial/\partial x_k$ denotes the filtered Lagrangian time derivative, $\widetilde{p}$ is the filtered pressure, ρ is the constant fluid mass density, ν is the constant kinematic viscosity, τ_{ij} is the subgrid-scale (SGS) stress tensor, and $\widetilde{S}_{ij} = (\partial \widetilde{U}_i/\partial x_j + \partial \widetilde{U}_j/\partial x_i)/2$ is the rate-of-strain tensor. The sum convention is used throughout this paper. The SGS stress is given by the quadratic model [2, 3]

$$\tau_{ij} = \frac{2}{3}k\delta_{ij} - 2\nu_t \widetilde{S}_{ij} - 3\frac{\nu_t^2}{k}\left[\widetilde{S}_{ik}\widetilde{\Omega}_{kj} + \widetilde{S}_{jk}\widetilde{\Omega}_{ki} - 2\widetilde{S}_{ik}\widetilde{S}_{kj} + \frac{2}{3}\widetilde{S}_{nk}\widetilde{S}_{kn}\delta_{ij}\right]. \tag{2}$$

Here, $k = \tau_{nn}/2$ refers to the SGS kinetic energy and $\widetilde{\Omega}_{ij} = (\partial \widetilde{U}_i/\partial x_j - \partial \widetilde{U}_j/\partial x_i)/2$ is the rate-of-rotation tensor. The SGS viscosity is given by $\nu_t = C_k L_m \sqrt{k}$, where C_k is a constant. The choice of the model length scale L_m decides about the model applied. This length scale is $L_m = \Delta$ if the dynamic model is used, and $L_m = \min(\Delta, L)$

if the unified model is applied, respectively. Here, L is defined by $L = k^{1.5}/\varepsilon$, and Δ is the filter width. The SGS kinetic energy and dissipation rate ε are calculated by

$$\frac{\widetilde{D}k}{\widetilde{D}t} = \frac{\partial}{\partial x_k}\left[(\nu+\nu_t)\frac{\partial k}{\partial x_k}\right] - \tau_{ij}\frac{\partial \widetilde{U}_j}{\partial x_i} - \frac{k^{3/2}}{L_m}, \tag{3}$$

$$\frac{\widetilde{D}\varepsilon}{\widetilde{D}t} = -C_1\frac{\varepsilon}{k}\tau_{ij}\frac{\partial \widetilde{U}_j}{\partial x_i} - C_2\frac{\varepsilon^2}{k} + \frac{\partial}{\partial x_k}\left[(\nu+\nu_t/\sigma_e)\frac{\partial k}{\partial x_k}\right]. \tag{4}$$

Here, $C_1 = 1.21, C_2 = 1.92$ and $\sigma_e = 1.3$ are usually applied constants.

The unified and dynamic models calculate the constants in different ways. Unified models apply $C_k = 1/9$. A linear unified model (LUM) neglects the quadratic term in Eq. (2), whereas a nonlinear unified model (NUM) applies Eq. (2). A linear dynamic model (LDM) neglects the quadratic term in Eq. (2) and calculates C_k dynamically. A nonlinear dynamic model (NDM) applies Eq. (2) and calculates the C_k values of linear and nonlinear terms separately by a dynamic method [3]. More details about these unified and dynamic methods can be found elsewhere [2, 3, 4, 5, 6].

3 Numerical Method and Problems Considered

Simulations were performed by using a finite-volume based method inside the OpenFOAM CFD Toolbox. The convection term in the momentum equation was discretized using a second-order central difference scheme. The PISO algorithm was used for the pressure-velocity coupling. The resulting algebraic equation for all the flow variables except pressure has been solved iteratively using a preconditioned bi conjugate gradient method with a diagonally incomplete LU preconditioning at each time step. The Poisson equation for the pressure was solved using an algebraic multi-grid (AMG) solver. When the scaled residual became less than 10^{-6}, the algebraic equation was considered to be converged. Time marching was performed using a second-order backward difference scheme. The time step was modified dynamically to ensure a constant CFL number of 0.5. Periodic boundary conditions have been employed along the streamwise and spanwise directions for all the flow variables. The wall-function approach was used to implement the boundary condition at the bottom of the domain while a stress free boundary condition was applied at the domain top. The consequence of using the wall-function approach is the need to replace negative values of the effective diffusivity in k and ε equations by zero if the dynamic approach is used.

Table 1 Set-up of the numerical simulations

Parameters	Wind tunnel	Neutral boundary layer
Domain	4.32 m $*$ 0.72 m $*$ 0.46 m	4 km $*$ 2 km $*$ 1.5 km
Grid	350 $*$ 80 $*$ 80	100 * 100 * 40
Roughness length (z_o)	0.03 mm	0.1 m

Two problems were considered, see the domain, grid resolution, and other parameters provided in Table 1. Regarding the wind tunnel simulations, the imposed pressure gradient drives the flow, whereas the geostrophic wind is used to drive the neutral boundary layer simulations. The motivation for performing wind tunnel simulations on fine grids is to show the suitability of models for atmospheric boundary layer (ABL) turbulence studies. But ABL LES is almost always performed on coarse grids. This motivates the application of turbulence models to neutral ABL simulations according to the benchmark problem of [7]. It should be noted that the grid resolution applied here is finer than the grids used in previous studies because a second-order central difference scheme is used here in the streamwise and spanwise directions, whereas spectral methods were used in previous studies. Systematic grid dependency studies are planned to analyze the performance of unified RANS-LES models in comparison with dynamic methods on coarse grids. We did not perform boundary condition sensitivity tests. Sullivan et. al [8] showed that lower boundary conditions using different wall-functions provide similar results.

4 Results and Discussion

The fine-grid comparisons of the mean streamwise velocity and Reynolds normal stress obtained from the dynamic and unified models considered with experimental data [9] are shown in Fig. 1. Results obtained by using the Lagrangian averaged scale dependent dynamic model (LASDD) [10] are also shown. The results obtained with the different models show a good agreement with each other and the experimental data. The NDM optimally predicts the peak of the Reynolds normal stress.

The coarse-grid comparisons of the normal components of the stress τ_{ij} obtained from the two unified and two dynamic models considered are shown in Fig. 2. For coarse grids, the SGS stress has to provide a main contribution to the total stress, which is known to be highly anisotropic. Hence, the SGS stress has to reflect this anisotropy. However, Fig. 2 shows that not all the models considered provide this anisotropy. The linear LUM and LDM predict isotropic stresses because they do not involve nonlinear terms. The nonlinear NUM and NDM predict an anisotropy of stresses. The anisotropy provided by the NUM is very small, whereas the prediction of the NDM shows characteristic features observed, e.g., for channel flow.

Another relevant difference between the nonlinear NUM and NDM can be seen in Fig. 3, which shows the vertical profiles of the normalized modeled, resolved, and total Reynolds shear stress. In LES, the resolved stress should continuously increase with the vertical height in correspondence to the behavior of the total stress. The NUM is not capable of simulating this shear stress behavior: the resolved stress has a discontinuity, and it is smaller than the modeled stress close to the surface. On the other hand, the NDM correctly describes this shear stress behavior: the resolved stress increases linearly with the height. The reason for the ability of the NDM to correctly describe the shear stress behavior is the inclusion of backscatter which tranfers kinetic energy from the modeled to the resolved scales [8].

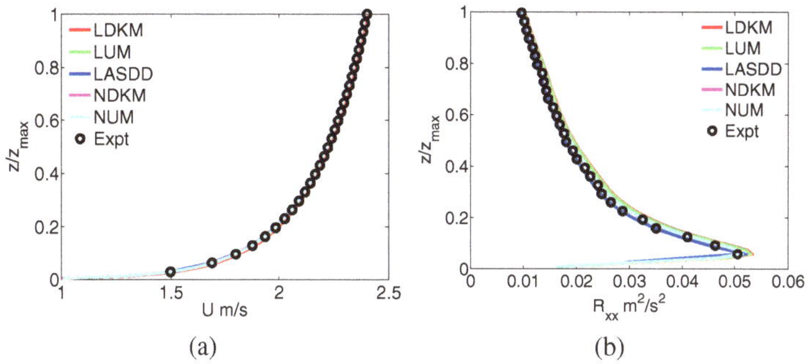

Fig. 1 Comparison of unified and dynamic models and the LASDD model with wind tunnel experiments [9]: (a) mean streamwise velocity, (b) streamwise Reynolds normal stress

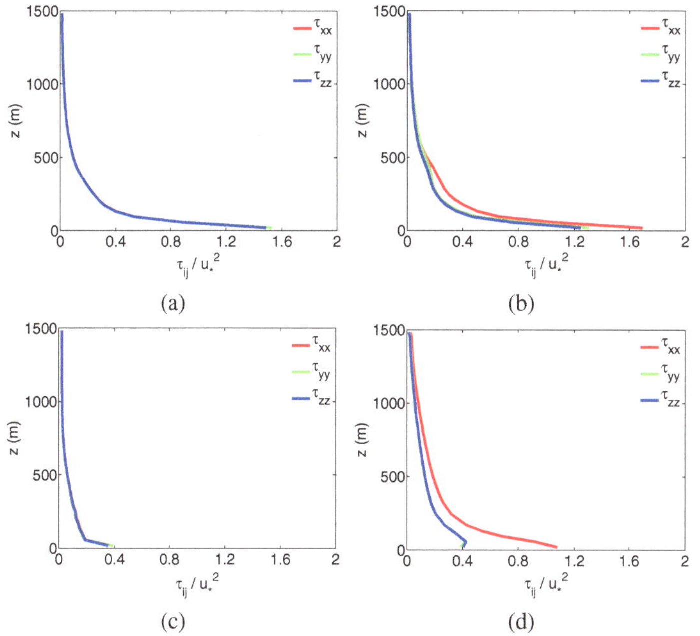

Fig. 2 Comparison of normalized modeled normal stresses obtained for the neutral ABL: (a) LUM, (b) NUM, (c) LDM, and (d) NDM. Here, u_* is the friction velocity and z is the vertical height.

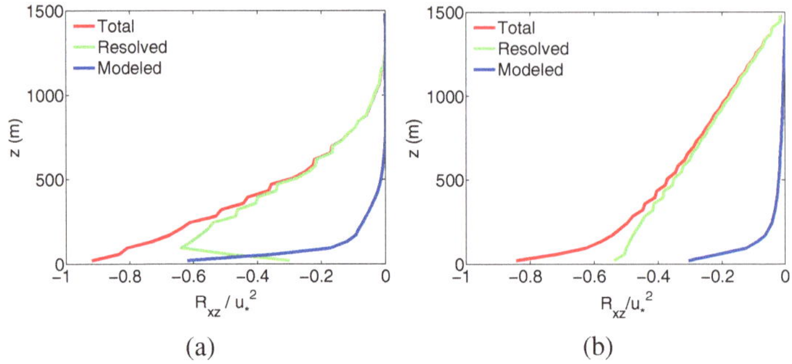

Fig. 3 Comparison of the normalized modeled, resolved, and total Reynolds shear stress: (a) NUM and (b) NDM. Here, u_* is the friction velocity and z is the vertical height.

5 Conclusions

The NDM model is found to be the most accurate model due to its ability to account for anisotropy and backscatter. Pictures of instantaneous streamwise velocities, which are not presented here, do also show that the NDM is capable of excellently representing the small-scale structure of turbulence. The NUM represents a valid alternative if computationally more efficient simulations have to be performed.

References

1. Dabiri, J.O.: Potential order-of-magnitude enhancement of wind farm power density via counter-rotating vertical-axis wind turbine arrays. IRESR 3(4), 043104/1–12 (2011)
2. Heinz, S.: Unified turbulence models for LES and RANS, FDF and PDF simulations. Theor. Comp. Fluid Dyn. 21(2), 99–118 (2007)
3. Heinz, S.: Realizability of dynamic subgrid-scale stress models via stochastic analysis. Monte Carlo Methods Appl. 14(4), 311–329 (2008)
4. Heinz, S., Gopalan, H., Stöllinger, M.: A unified RANS-LES model. Part 1. Computational model development. J. Comput. Phys. (2012)
5. Gopalan, H., Heinz, S., Stöllinger, M.: A unified RANS-LES model. Part 2. Model accuracy and computational cost. J. Comput. Phys. (2012)
6. Heinz, S., Gopalan, H.: Realizable versus non-realizable dynamic subgrid-scale stress models. Phys. Fluids 24 (2012) (submitted)
7. Andren, A., Brown, A.R., Mason, P.J., Graf, J., Schumann, U., Moeng, C.H., Nieuwstadt, F.T.M.: Large-eddy simulation of a neutrally stratified boundary layer: A comparison of four computer codes. Q. J. Roy. Meteo. Soc. 12(520), 1457–1484 (1994)
8. Sullivan, P.P., McWilliams, J.C., Moeng, C.H.: A subgrid-scale model for large-eddy simulation of planetary boundary-layer flows. Bound-Lay. Meteorol. 71(3), 247–276 (1994)
9. Chamorro, L.P., Porté-Agel, F.: A wind-tunnel investigation of wind-turbine wakes: boundary-layer turbulence effects. Bound-Lay. Meteorol. 132(1), 129–149 (2009)
10. Wu, Y.T., Porté-Agel, F.: Large-eddy simulation of wind-turbine wakes: evaluation of turbine parametrisations. Bound-Lay. Meteorol. 138(3), 345–366 (2011)

Large-Eddy Simulations of Wind Turbine Dedicated Airfoils at High Reynolds Numbers

O. Lehmkuhl, J. Calafell, I. Rodríguez, and A. Oliva

Abstract. This work aims at modelling the flow behaviour past airfoils used for wind turbine blades at high Reynolds number and large AoA. Three profiles have been selected: DU-93-W-210, DU-91-W2-250 and FX-77-W-500. To do this, a parallel unstructured conservative formulation has been used together with wall-adapting local-eddy viscosity model within a variational multi-scale framework (VMS-WALE). A methodology based on unstructured meshes has been developed in order to optimize the performance of the grid for LES calculations. Numerical results are presented in comparison with experimental ones for each airfoil. A good agreement between them can be observed.

1 Introduction

The flow around aerodynamic profiles in pre- or full-stall at high Reynolds numbers is a problem of increasing interest since it is a normal operation state for wind turbine blades. In the past years, it has been subject of many experimental and numerical investigations. Most of the numerical studies performed since now have been carried out using RANS modelling, but it is well-known that such models fails in predicting the flow at angles of attack (AoA) near or after stall, mainly due to the highly unsteady and three-dimensionality nature of the flow. Under these situations, large-eddy simulationes (LES) can be a good alternative for simulating such complex flows.

This work aims at modelling the flow behaviour past airfoils used for wind turbine blades at high Reynolds number and large AoA. Three profiles have been

O. Lehmkuhl · J. Calafell
Technical University of Catalonia, Spain and Termo Fluids S.L, Spain
e-mail: {oriol,joancs}@cttc.upc.edu

I. Rodríguez · A. Oliva
Technical University of Catalonia, Spain
e-mail: {ivette,oliva}@cttc.upc.edu

M. Hölling et al. (eds.), *Wind Energy – Impact of Turbulence*,
Research Topics in Wind Energy 2,
DOI: 10.1007/978-3-642-54696-9_22,

selected: DU-93-W-210, DU-91-W2-250 and FX77-W-500. To do this, a parallel unstructured conservative formulation has been used. The wall-adapting local-eddy viscosity model within a variational multi-scale framework (VMS-WALE) has been used to model subgrid scales (SGS) in the present study. This model has been proved to perform well on unstructured grids [5]. The study has been carried out for Reynolds numbers up to 3×10^6 and AoA up to 16°. Numerical results have been compared with experimental ones. In addition, in order to assess the performance of the LES, an *a-posteriori* methodology, which quantify the ratio of the solved scales to the smallest scales of the flow has been used.

2 Mathematical and Numerical Model

The turbulent flow is described by means of LES using symmetry-preserving discretizations. The spatial filtered and discretized Navier-Stokes equations can be written as,

$$\mathsf{M}u = 0 \tag{1}$$

$$\Omega\frac{\partial \overline{u}}{\partial t} + \mathsf{C}(\overline{u})\overline{u} + \nu\mathsf{D}\overline{u} + \rho^{-1}\Omega\mathsf{G}\overline{p} \;=\; \mathsf{C}(\overline{u})\overline{u} - \overline{\mathsf{C}(u)u} \;\approx\; -\mathscr{M}\mathscr{T}_m \tag{2}$$

where M, C, D and G are the divergence, convective, diffusive and gradient operators, respectively, Ω is a diagonal matrix with the sizes of control volumes, ρ is the fluid density, ν the viscosity, $\overline{p}$ represents the filtered pressure, $\overline{u}$ is the filtered velocity, $\mathscr{M}$ represents the divergence operator of a tensor, and $\mathscr{T}_m$ is the SGS stress tensor.

To close the formulation, a suitable expression for the SGS viscosity, ν_{sgs}, must be introduced. In the present work the WALE model [6] within a variational multiscale framework [3] (VMS-WALE) has been used. In the variational multiscale (VMS) approach, originally formulated for the Smagorinsky model by Hughes in the Fourier space, three classes of scales are considered: large, small and unresolved scales. The SGS stress tensor is only calculated for the small scales ($\mathscr{T}_m \approx \mathscr{T}'$).

In our implementation the *small-small* strategy is used in conjunction with the WALE model:

$$\mathscr{T}' = -2\nu_{sgs}\mathscr{S}_{\mathrm{ij}}' + \frac{1}{3}\mathscr{T}'\delta_{ij} \tag{3}$$

$$\nu_{sgs} = (C_w^{vms}\Delta)^2 \frac{(\mathscr{V}_{\mathrm{ij}}' : \mathscr{V}_{\mathrm{ij}}')^{\frac{3}{2}}}{(\mathscr{S}_{\mathrm{ij}}' : \mathscr{S}_{\mathrm{ij}}')^{\frac{5}{2}} + (\mathscr{V}_{\mathrm{ij}}' : \mathscr{V}_{\mathrm{ij}}')^{\frac{5}{4}}}$$

$$\mathscr{S}_{\mathrm{ij}}' = \frac{1}{2}[\mathsf{G}(\overline{u'}) + \mathsf{G}^*(\overline{u'})]$$

$$\mathscr{V}_{\mathrm{ij}}' = \frac{1}{2}[\mathsf{G}(\overline{u'})^2 + \mathsf{G}^*(\overline{u'})^2] - \frac{1}{3}(\mathsf{G}(\overline{u'})^2\mathsf{I})$$

where C_w^{vms} is the equivalent of the WALE coefficient for the *small-small* VMS approach and in the finite volume context its value lies in the range between 0.3 and 0.5. In our studies a value of 0.325 is used.

The governing equations have been discretized on a collocated unstructured grid arrangement by means of second-order spectro-consistent schemes [9]. Such schemes are conservative, i.e. they preserve the symmetry properties of the continuous differential operators and ensure both, stability and conservation of the kinetic-energy balance even at high Reynolds numbers and with coarse grids. These conservation properties are held if, and only if the discrete convective operator is skew-symmetric ($\mathsf{C}(u) = -\mathsf{C}^*(u)$), the negative conjugate transpose of the discrete gradient operator is exactly equal to the divergence operator ($-(\Omega\mathsf{G})^* = \mathsf{M}$) and the diffusive operator D, is symmetric and positive-definite. For the temporal discretisation of the momentum equation a two-step linear explicit scheme on a fractional-step method has been used for the convective and diffusive terms, while for the pressure gradient term an implicit first-order scheme has been used. This methodology has been previously used with accurate results for solving the flow over bluff bodies with massive separation [4, 7].

3 Assessment of Mesh Resolution

Numerical simulations have been performed on different grids (about 3-23 million control volumes). In order to asses the grid resolution quality, the ratio of the resolved scales to the smallest ones of the real flow has been evaluated:

$$\Gamma = \frac{h}{\eta} = \frac{{V_{cell}}^{1/3}}{\left(\nu^3/\varepsilon\right)^{1/4}} \quad \text{where} \quad \varepsilon = 2\left(\nu + \nu_{sgs}\right)\overline{\mathscr{S}_{\mathrm{ij}}{}' \mathscr{S}_{\mathrm{ij}}{}'} \tag{4}$$

This ratio gives an orientation of the contribution of the LES model to dissipation and therefore a quantification of the modelled dissipation with respect to the resolved one. Therefore, in order to obtain credible results, special care is taken during the mesh generation process, so as to maintain this ratio within reasonable limits.

In figure 1 the ratio distribution is shown for the FX77-W-500 case. A preliminar calculation allows us to assess the initial map. These results are used to improve the inital mesh by adding control volumes in the low resolved areas making the ratio distribution as uniform as possible. Then a second and final computation is performed.

4 Illustrative Results

Three airfoil profiles have been simulated, DU-91-W2-210, DU-93-W-250 and FX77-W-500. For each airfoil, LES results have been compared with experimental data provided by Delft University of Technology (DUT) [8], Risø National Laboratory (Denmark) [2] and the Institut für Aerodynamik und Gasdynamik (Germany)

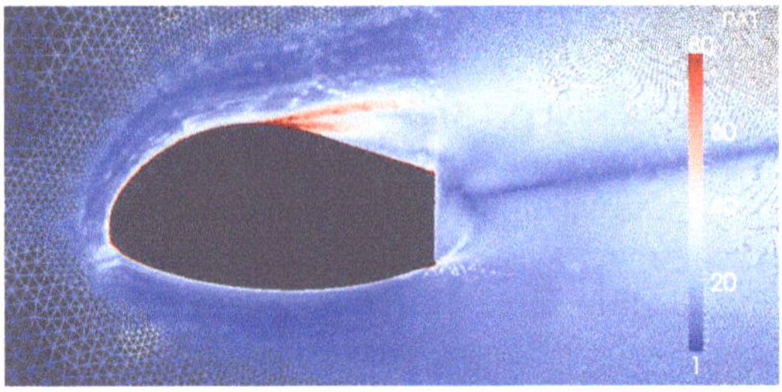

Fig. 1 Ratio of the resolved scales to the smallest ones for the used FX77-W-500 mesh at $Re = 2.75 \times 10^6$

[1], respectively. Results about lift and drag coefficients are plotted in figure 2 for the airfoil DU-91-W2-210 at $Re = 3 \times 10^6$. In addition, in figure 3, lift coefficient results for DU-93-W-250 at $Re = 1 \times 10^6$ and FX77-W-500 at $Re = 2.75 \times 10^6$ are depicted. Good agreement between the calculated results and the experimental ones is achieved, specially at angles of attack (AoA) with important boundary layer deattachment i.e. near stall and post-stall. Some illustrative instantaneous results are depicted. In figure 4 the FX77-W-500 airfoil with attached boundary layer along the upper surface is shown and in figure 5 a strong flow separation is displayed.

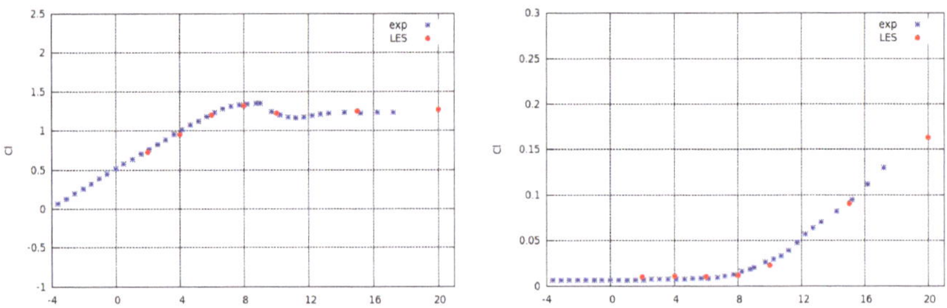

(a) (b)

Fig. 2 Aerodynamic coefficients at $Re = 3 \times 10^6$ and different AoA for DU-93-W-210 airfoil. (a) Lift coefficient (b) Drag coefficient.

As a result of several calculations with different meshes, it has been observed that having an appropriate mesh is of paramount importance to make the subgrid model work properly. The main features that a mesh should have from the geometric point of view, is a proper aspect ratio of the control volumes near the airfoil and smooth size transitions between neighbouring cells. Regarding the mesh resolution, a uniform distribution of the ratio of the resolved scales to the smallest ones is also important while keeping those values whithin a reasonable range. In order to meet all the previous requierements, it is almost unavoidable the use of ustructured meshes, that allow us to adapt the mesh size locally and optimizing the distribution of control volumes placing them where they are really needed. The combination of the VMS-WALE LES model together with an optimized mesh lead us to the present results that reproduces fairly well the physical phenomena. Instantaneous

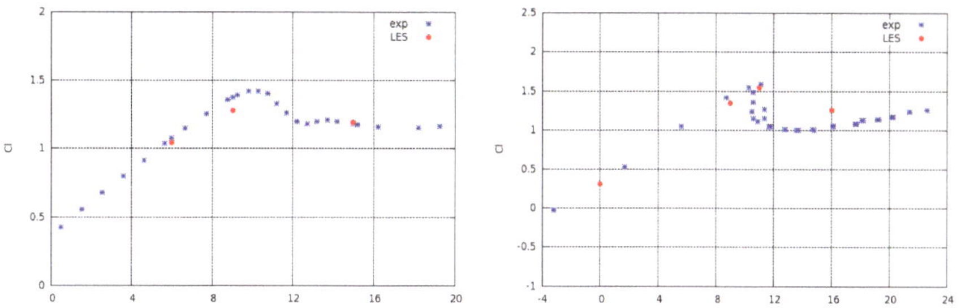

(a) (b)

Fig. 3 Lift coefficient as a function of the AoA (a) DU-91-W2-250 at $Re = 1 \times 10^6$ (b) FX77-W-500 at $Re = 2.75 \times 10^6$

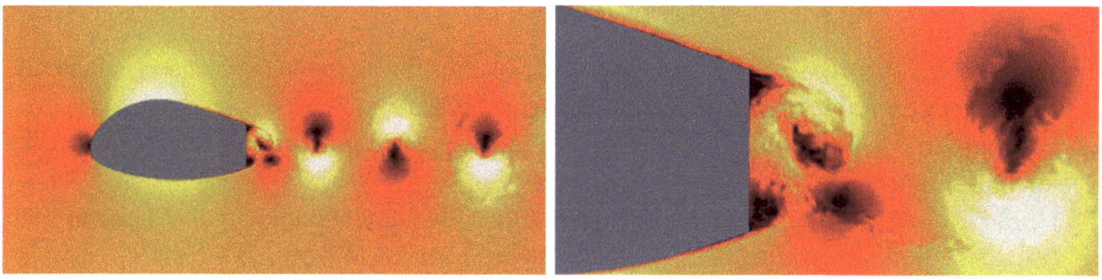

(a) (b)

Fig. 4 Illustrative results. Velocity contours for FX77-W-500 airfoil at AoA = 0.

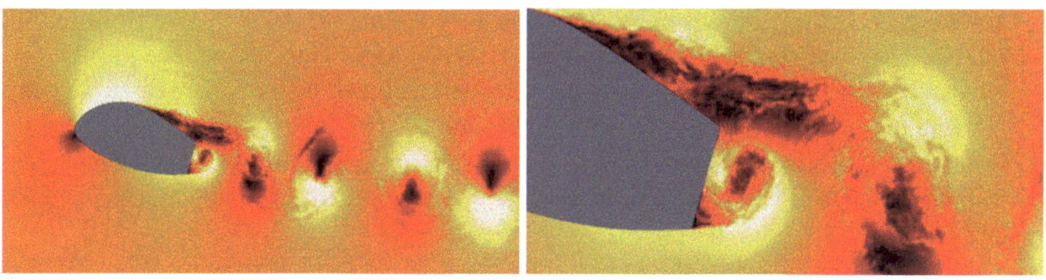

(a) (b)

Fig. 5 Illustrative results. Velocity contours for FX77-W-500 airfoil at AoA = 16.

flow captures display a large number of flow scales which are well represented by the model.

5 Conclusions

In this paper LES calculations have been performed to determine the aerodynamic coefficients of the DU-91-W2-210, DU-93-W-250 and FX77-W-500 airfoils. The wall-adapting local-eddy viscosity model within a variational multi-scale framework (VMS-WALE) has been used as the SGS model. It has been observed that in order to obtain a good agreement between numerical LES results and experimental ones not only the mathematical formulation is important but also the mesh performance. By combining properly both requierements in the way reported in the present paper, good results are obtained.

Acknowledgements. The authors thank the financial support provided by EIT through KIC InnoEnergy Offshore Wind Enabling Technology "OFFWINDTECH" project. This work has been also partially financially supported by the Ministerio de Ciencia e Innovacin, Spain (Project ENE-2010-17801) and by a collaboration project between Universitat Politècnica de Catalunya and Termo Fluids S.L. (ref. C06650). The authors would like to acknowledge sincerely W.A. Timmer and Dr.-Ing. Thorsten Lutz for providing their experimental data.

References

1. Althaus, D.: Niedriggeschwindigkeitsprofile. Institut für Aerodynamik und Gasdynamik, Luftfahrzeugaerodynamik. Universität Stuttgart (1980)
2. Bertagnolio, F., Sørensen, N., Johansen, J., Fuglsang, P.: Wind Turbine Airfoil Catalogue. Technical report (2001)
3. Hughes, T.J.R., Mazzei, L., Jansen, K.E.: Large eddy simulation and the variational multiscale method. Computing and Visualization in Science 3, 47–59 (2000)
4. Lehmkuhl, O., Borrell, R., Chiva, J., Perez-Segarra, C.D.: Direct numerical simulations and symmetry-preserving regularization simulations of the flow over a circular cylinder at Reynolds number 3900. Turbulence, Heat and Mass Transfer (2009)
5. Lehmkuhl, O., Baez, A., Rodríguez, I., Perez-Segarra, C.D.: Direct numerical simulation and Large-Eddy simulations of the turbulent flow around a NACA-0012 airfoil. In: 7th International Conference on Computational Heat and Mass Transfer, pp. 1–8 (2011)
6. Nicoud, F., Ducros, F.: Subgrid-scale stress modelling based on the square of the velocity gradient tensor. Flow, Turbulence and Combustion 62, 183–200 (1999)
7. Rodríguez, I., Borrell, R., Lehmkuhl, O., Pérez-Segarra, C.D., Oliva, A.: Direct Numerical Simulation of the Flow Over a Sphere at Re = 3700. Journal of Fluids Mechanics 679, 263–287 (2011)
8. Timmer, W.A., Van Rooij, R.P.J.O.M.: Summary of the Delft University Wind Turbine Dedicated Airfoils. Journal of Solar Engineering 125(4), 1–11 (2004)
9. Verstappen, R.W.C.P., Veldman, A.E.P.: Symmetry-Preserving Discretization of Turbulent Flow. Journal of Computational Physics 187, 343–368 (2003)

Low-Order Modelling of Blade-Induced Turbulence for RANS Actuator Disk Computations of Wind and Tidal Turbines

Takafumi Nishino and Richard H.J. Willden

Abstract. Modelling of turbine blade-induced turbulence (BIT) is discussed within the framework of three-dimensional Reynolds-averaged Navier-Stokes (RANS) actuator disk computations. We first propose a generic (baseline) BIT model, which is applied only to the actuator disk surface, does not include any model coefficients (other than those used in the original RANS turbulence model) and is expected to be valid in the limiting case where BIT is fully isotropic and in energy equilibrium. The baseline model is then combined with correction functions applied to the region behind the disk to account for the effect of rotor tip vortices causing a mismatch of Reynolds shear stress between short- and long-time averaged flow fields. Results are compared with wake measurements of a two-bladed wind turbine model of Medici and Alfredsson [Wind Energy, Vol. 9, 2006, pp. 219-236] to demonstrate the capability of the new model.

1 Introduction

Modelling of turbulent mixing behind turbines is an important area of research for the wind and tidal power industries. Whilst recent eddy-resolving computations (such as large-eddy simulations and detached-eddy simulations) using a so-called actuator line method have given valuable insight into time-dependent features of wind/tidal turbine wakes (e.g., [1]), it is still of great importance to develop a low-order model that satisfactorily predicts the characteristics of time-averaged flow around turbines at much lower computational cost. Such a low-order model would be useful for the design of wind and tidal power farms for the future.

The objective of the present study is to develop a reasonably generic low-order model that describes time-averaged flow around a horizontal-axis wind/tidal

Takafumi Nishino · Richard H.J. Willden
Department of Engineering Science, University of Oxford, Oxford, United Kingdom
e-mail: {takafumi.nishino,richard.willden}@eng.ox.ac.uk

M. Hölling et al. (eds.), *Wind Energy – Impact of Turbulence*,
Research Topics in Wind Energy 2,
DOI: 10.1007/978-3-642-54696-9_23,

turbine of various designs and operating conditions. Our current model is based on the three-dimensional Reynolds-averaged Navier Stokes (RANS) equations coupled with an actuator disk model. To account for the effect of turbulence (or velocity fluctuations) induced by turbine blades—we refer to this as *blade-induced turbulence* (BIT) in this paper—we first introduce a simple baseline model, which requires the specification of BIT energy and scale but does not require any additional model coefficients (other than those used in the original RANS turbulence model). The baseline model is then combined with correction functions to account for the effect of rotor tip vortices. Details of the model are described below, followed by the results of a model validation study.

2 Baseline Model

The baseline model used in this study is the one recently proposed by the authors [5]. The governing equations of the flow are the three-dimensional incompressible RANS equations, where the Reynolds stress tensor, $-\overline{u_i'u_j'}$, is modelled using the standard k-ε model of Launder and Spalding [2].

A simple actuator disk concept is used, where a turbine rotor is modelled as a stationary permeable disk of zero thickness placed perpendicular to the incoming flow. The effect of the rotor on the mean flow is considered as a loss of streamwise (or axial) momentum at the disk plane. The change in momentum flux (per unit disk-area and per unit fluid-density) is locally calculated as $S_U = \frac{1}{2}KU_d^2$, where U_d is the local streamwise velocity at the disk plane and K is a momentum loss factor (constant over the disk) to determine the thrust acting on the disk.

It is known that this actuator disk model (without taking account of BIT) yields much weaker turbulent mixing behind the disk compared to that measured behind a turbine rotor. In the baseline BIT model we assume that, at the disk plane, virtual turbine blades generate turbulence that is characterised by its turbulent kinetic energy k_b and dissipation rate ε_b. Two physical parameters are introduced to determine the values of k_b and ε_b: (i) the ratio of the energy converted to BIT to that removed from the mean flow at the disk plane, β, and (ii) a representative length scale for BIT, l_b. Note that β and l_b are model variables rather than model coefficients, as they depend on actual turbine design and operating conditions. k_b is then calculated (locally over the disk plane) as $k_b = \beta S_U = \frac{1}{2}\beta KU_d^2$, whereas ε_b is estimated (based on the high Reynolds number equilibrium hypothesis) as $\varepsilon_b = C_\mu^{\frac{3}{4}} k_b^{\frac{3}{2}} / l_b$, where $C_\mu = 0.09$ (following the standard k-ε model).

To account for the combined effect of BIT and the turbulence coming from upstream of the turbine, we further assume that: (i) the disk plane is a special internal boundary that does not allow the transport of k or ε via their diffusion (so that the transport of k and ε through the disk plane is only via the convection from upstream to downstream) and (ii) at the disk plane, the BIT of k_b and ε_b is mixed (in the time-averaged sense) with the *ambient* (or upstream) turbulence of k_a and ε_a, resulting in the *mixed* turbulence of k_m and ε_m just downstream of the disk. Here k_m is calculated as $k_m = k_a + k_b$, whereas ε_m is estimated by assuming the conservation of the time-integral of linearly decaying turbulent kinetic energy, i.e., $k_m \tau_m = k_a \tau_a + k_b \tau_b$, where $\tau_a = k_a/\varepsilon_a$, $\tau_b = k_b/\varepsilon_b$ and $\tau_m = k_m/\varepsilon_m$ represent the initial eddy turnover time or lifetime for the ambient, blade-induced and mixed turbulence, respectively [5]. Eventually, the changes in k and ε to be added to their transport equations at the disk plane (per unit disk-area) are calculated as

$$S_k = U_d(k_m - k_a) = U_d k_b \ , \tag{1}$$

$$S_\varepsilon = U_d(\varepsilon_m - \varepsilon_a) = U_d \left[\frac{(k_a + k_b)^2}{(k_a^2/\varepsilon_a) + (k_b^2/\varepsilon_b)} - \varepsilon_a \right] . \tag{2}$$

Generally, the two model variables, β and l_b, should depend on the actual turbine design and may be given as functions of the distance, r, from the disk axis. In the present study, however, uniform values of β and l_b are given either to the entire disk surface or to the disk edge region defined by $(0.5d - w_{\text{edge}}) \leq r \leq 0.5d$, where d is the disk diameter and w_{edge} ($= 0.1d$ in this study) is the width of the disk edge region. For the latter case, $\beta = 0$ is given to the rest of the disk surface.

3 Tip Vortex Correction

As will be shown later, the baseline model tends to yield too strong/fast turbulent mixing in the core region ($r/d < 0.4$) in the near wake, even when BIT is given only around the disk edge. This suggests that further modifications are required on the modelling of turbulent mixing in the near wake region.

An important issue to be considered here is how large the Reynolds shear stress is (relative to the turbulent kinetic energy) behind a turbine, especially in the region where strong rotor tip vortices exist. It is known that for many two-dimensional turbulent shear flows where the production and dissipation of k are close to equilibrium, the value of $\overline{u_x' u_y'}/k$ is around 0.3 (when the mean shear $\mathrm{d}U/\mathrm{d}y < 0$) or -0.3 (when $\mathrm{d}U/\mathrm{d}y > 0$) and this is the basis of the turbulent viscosity constant $C_\mu = (0.3)^2 = 0.09$ used in the standard k-ε model (see, e.g., Pope [6]). In the mean shear layer behind a turbine rotor, however, experimental results have shown that the magnitude of $\overline{u_x' u_y'}/k$ can be significantly smaller than 0.3 [4]. This is most likely due to the effect of tip vortices causing a mismatch between short-time (or phase) averaged and long-time averaged wake shear profiles. Specifically, tip vortices periodically create "adverse" velocity gradient regions, $\partial \langle u \rangle / \partial r < 0$ (where $\langle \phi \rangle$ denotes a short-time average of ϕ over a time scale relevant to the passing-through time of each tip vortex), and hence "adverse" Reynolds shear stress regions, $\langle u_x' u_r' \rangle > 0$, whilst the long-time averaged wake shear direction is $\partial U/\partial r > 0$ and therefore $\overline{u_x' u_r'} < 0$. This explains why the long-time-averaged wake shear behind a rotor does not diffuse quickly despite its high level of turbulent kinetic energy.

To account for the above effect of tip vortices within the framework of RANS actuator disk computations, we consider applying empirical correction functions to the so-called tip vortex region. The streamwise extent of the region will be given as a model variable, l_{tip}, which may be further modelled as a function of turbine design and operating conditions in future studies. For the present study, we introduce three correction functions, f_μ, f_k and $f_{\varepsilon 1}$, into the standard k-ε model:

$$\frac{Dk}{Dt} = \frac{\partial}{\partial x_j}\left(\frac{\nu_t}{\sigma_k}\frac{\partial k}{\partial x_j}\right) - \overline{u'_i u'_j}\frac{\partial U_i}{\partial x_j} f_k - \varepsilon \,, \tag{3}$$

$$\frac{D\varepsilon}{Dt} = \frac{\partial}{\partial x_j}\left(\frac{\nu_t}{\sigma_\varepsilon}\frac{\partial \varepsilon}{\partial x_j}\right) - \overline{u'_i u'_j}\frac{\partial U_i}{\partial x_j} f_{\varepsilon 1} C_{\varepsilon 1}\frac{\varepsilon}{k} - C_{\varepsilon 2}\frac{\varepsilon^2}{k} \,, \tag{4}$$

$$-\overline{u'_i u'_j} = \nu_t\left(\frac{\partial U_i}{\partial x_j} + \frac{\partial U_j}{\partial x_i}\right) - \frac{2}{3}k\delta_{ij} \,, \quad \nu_t = f_\mu C_\mu \frac{k^2}{\varepsilon} \,, \tag{5}$$

where $C_\mu = 0.09, C_{\varepsilon 1} = 1.44, C_{\varepsilon 2} = 1.92, \sigma_k = 1.0$ and $\sigma_\varepsilon = 1.3$ (following the standard k-ε model). To reduce the eddy viscosity and thus mitigate turbulent mixing in the tip vortex region behind the disk (located at $x = 0$), we model f_μ as follows:

$$f_\mu = 1 \quad \text{for} \quad x \leq 0 \,,\ l_{\text{tip}} \leq x \,, \tag{6}$$

$$f_\mu = 1 - f_r(1 - f_{\mu\text{min}}) \quad \text{for} \quad 0 \leq x \leq l^*_{\text{tip}} \,, \tag{7}$$

$$f_\mu = 1 - f_r(1 - f_{\mu\text{min}})\frac{1}{2}\left(1 + \cos\frac{\pi(x - l^*_{\text{tip}})}{l_{\text{tip}} - l^*_{\text{tip}}}\right) \quad \text{for} \quad l^*_{\text{tip}} \leq x \leq l_{\text{tip}} \,, \tag{8}$$

where $f_{\mu\text{min}} = 0.1$ is given in this study, following experimental observations [4] (note, however, that this is also a model variable and should generally depend on the rotor tip-speed ratio and the number of blades). l^*_{tip} is the streamwise extent of the region where f_μ does not change (corresponding to the region where tip vortices are stable); we assume $l^*_{\text{tip}} = \frac{1}{2}l_{\text{tip}}$ in the present model. f_r is a damping (sinusoidal) function in the radial (r) direction, normalised such that $f_r = 1$ at $r = 0.5d$ (where the rotor edge is located) and monotonically decreases to $f_r = 0$ at $r = 0$ and d.

A major difficulty in this correction is that the change in ν_t due to the introduction of f_μ affects not only the strength of turbulent diffusion but also the production of k and ε. This necessitates the introduction of f_k and $f_{\varepsilon 1}$ in Eqs (3) and (4), respectively; however, we assume $f_{\varepsilon 1} = 1$ in this study for simplicity.[1] Preliminary computations with $f_k = 1$ yielded too small k in the near wake, whereas those with $f_k = 1/f_\mu$ (i.e., fully cancelling out the effect of f_μ on the production of k) resulted

[1] Here the introduction of f_k is justified since, in the tip vortex region, the production of k should be linked to the mean of the magnitude of short-time averaged velocity gradient and the corresponding Reynolds stress rather than those for the long-time averaged flow field, i.e., f_k accounts for the difference between $-\overline{u'_i u'_j}\frac{\partial U_i}{\partial x_j}$ (known in the model) and $\overline{|-\langle u'_i u'_j\rangle \frac{\partial \langle u_i\rangle}{\partial x_j}|}$ (unknown in the model). Meanwhile, physical interpretation of $f_{\varepsilon 1}$ seems less straightforward since the production term of ε includes ε and k, which should also be divided into two components corresponding to the short-time and long-time averaged flow fields, respectively. This seems far beyond the capability of single-scale eddy viscosity models (perhaps we require a proper multi-scale model for such discussion; see, e.g., Sagaut et al. [7]) and hence we assume $f_{\varepsilon 1} = 1$ in this study.

in too large k in the near wake. To maintain the level of turbulent kinetic energy in the near wake comparable to that measured in the experiments [4] (and also to minimise its sensitivity to the value of l_{tip}), we model f_k as $f_k = 1 + \gamma_k(1 - f_\mu)/f_\mu$ with the model coefficient $\gamma_k = 0.85$ in this study (note that $\gamma_k = 0$ and 1 correspond to the two extreme cases, $f_k = 1$ and $1/f_\mu$, respectively).

4 Model Validation

A model validation study was performed for wind tunnel tests of a two-bladed wind turbine model of Medici and Alfredsson [3, 4]. The tunnel test section is 1.2m wide and 0.8m high. The tunnel ceiling is adjustable, so that the freestream velocity, U_{in}, is constant throughout the test section (when the turbine is not installed). The rotor diameter d is 0.18m and its hub height from the floor is 0.24m. The hub diameter is 20mm. Table 1 summarises experimental conditions; for this particular set of experiments, a turbulence-generating grid was installed, providing freestream turbulence (FST) intensities of 4.5% at $x/d = 0$ (turbine location) and 2.5% at $x/d = 9$. Here the thrust coefficient C_T and the tip-speed ratio (TSR) are based on the freestream velocity and therefore slightly different from those reported earlier [3].

For the computations, the turbine rotor was modelled as an actuator disk of zero thickness with a small non-permeable disk embedded at the centre to account for the effect of the rotor hub. The supporting tower of the turbine was not modelled. Computations were performed using FLUENT 12 (together with its User-Defined Function module). The solver is based on a finite volume method and is nominally 2nd-order accurate in space. All computations were performed as steady state.

Table 1 Experimental conditions

U_{in}	TSR	C_T	Yaw angle	FST intensity
8.14 [m/s]	3.87	0.899	0 [deg.]	4.5% ($x/d = 0$), 2.5% ($x/d = 9$)

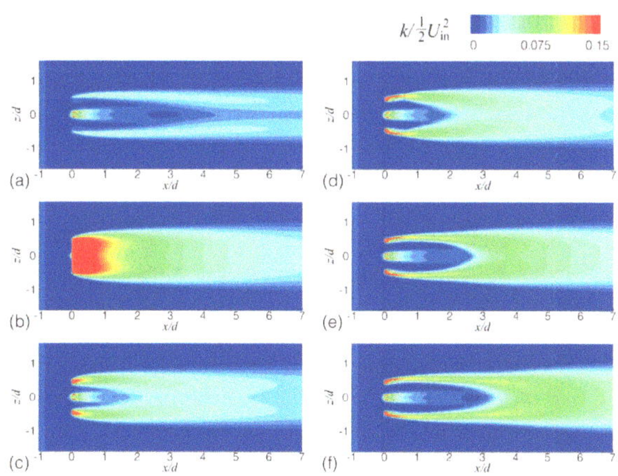

Fig. 1 Contours of turbulent kinetic energy across a horizontal plane at hub height: (a) without BIT, (b) baseline model (for the entire actuator disk), (c) baseline model (only for $0.4 < r/d < 0.5$), and (d-f) baseline model (only for $0.4 < r/d < 0.5$) with the tip vortex correction ($l_{\text{tip}}/d = 1$, 3, 5 for d, e, f, respectively)

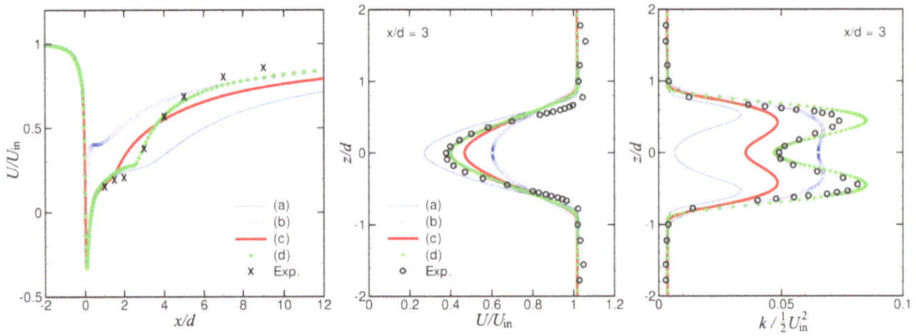

Fig. 2 Centreline streamwise velocity (left), streamwise velocity profiles at $x/d = 3$ (centre) and turbulent kinetic energy profiles at $x/d = 3$ (right): (a) without BIT, (b) baseline model (for the entire actuator disk), (c) baseline model (only for $0.4 < r/d < 0.5$), (d) baseline model (only for $0.4 < r/d < 0.5$) with the tip vortex correction ($l_{\text{tip}}/d = 3$), and experimental data [3]

Figs. 1 and 2 present results for $C_T = 0.899$, $\beta = 0.30$ and $l_b = 0.1d$ (except for the case without BIT). The case without BIT yields too small turbulent kinetic energy and thus too slow wake recovery. The baseline model applied to the entire disk yields too large turbulent kinetic energy in the core region ($r/d < 0.4$) and hence too early wake recovery behind the disk. The baseline model applied only to the edge region still yields too strong turbulent mixing in the core region in the near wake, and the tip vortex correction provides better results in both near and far wake. Further validation studies are needed to examine the robustness of the model.

Acknowledgements. The authors gratefully acknowledge the support of the Oxford Martin School, University of Oxford, who have funded this research. The authors would also like to thank Dr Davide Medici and Prof. Henrik Alfredsson for providing us with their experimental data.

References

1. Ivanell, S., Srensen, J.N., Mikkelsen, R., Henningson, D.: Analysis of numerically generated wake structures. Wind Energy 12, 63–80 (2009)
2. Launder, B.E., Spalding, D.B.: The numerical computation of turbulent flows. Comput. Method. Appl. M. 3, 269–289 (1974)
3. Medici, D., Alfredsson, P.H.: Measurements on a wind turbine wake: 3D effects and bluff body vortex shedding. Wind Energy 9, 219–236 (2006)
4. Medici, D., Alfredsson, P.H.: Personal communication (2011)
5. Nishino, T., Willden, R.H.J.: Effects of 3-D channel blockage and turbulent wake mixing on the limit of power extraction by tidal turbines. Int. J. Heat Fluid Flow (under review)
6. Pope, S.B.: Turbulent Flows. Cambridge University Press, Cambridge (2000)
7. Sagaut, P., Deck, S., Terracol, M.: Multiscale and Multiresolution Approaches in Turbulence. Imperial College Press, London (2006)

On LES Assessment in Massive Separated Flows: Flow Past a NACA Airfoil at Re = 50000

I. Rodríguez, O. Lehmkuhl, A. Baez, and C.D. Perez-Segarra

Abstract. Mechanisms of quasi-periodic flow oscillation observed in airfoils near stall and stall behaviour, which affect airfoil efficiency, remain still not fully understood. In this sense, numerical simulations using subgrid-scale models can be useful in order to shed some light in the underlying complex physics. However, the assessment of these models is an important aspect to consider for obtaining credible results. In this work, two LES models are compared with results from direct numerical simulation. The results have shown that LES models are capable of capturing the complex flow phenomena present at these AOA such as the low frequency fluctuations near stall and the vortex shedding produced at higher AOAs.

1 Introduction

Stall on airfoils is caused by massive separation of the flow leading to a sharp drop in the lift and an increase in the drag over the airfoil surface. NACA0012 airfoil exhibits two types of stall. A trailing-edge stall at all Reynolds numbers and a combined leading-edge/trailing-edge stall at intermediate Reynolds number. The latter shows the combined presence at stall of a turbulent boundary layer separation moving forward from the trailing-edge and a small laminar bubble in the leading-edge region failing to reattach which complete the flow breakdown. An oscillating situation

I. Rodríguez · A. Baez · C.D. Perez-Segarra
Heat and Mass Transfer Technological Center (CTTC), Technical University of Catalonia, Terrassa, Spain
e-mail: ivette@cttc.upc.edu

O. Lehmkuhl
TermoFluids S.L.
e-mail: oriol@termofluids.com
University of Catalonia
e-mail: aleix@cttc.upc.edu
(CTTC), Technical University of Catalonia
e-mail: segarra@cttc.upc.edu

M. Hölling et al. (eds.), *Wind Energy – Impact of Turbulence*,
Research Topics in Wind Energy 2,
DOI: 10.1007/978-3-642-54696-9_24,

is often noticed near stall angles. However, mechanisms of quasi-periodic oscillation observed near stall and stall behaviour, which affect airfoil efficiency, remain still not fully understood. In this sense, Large-Eddy Simulations (LES) have been performed. Two different models have been considered, the Wall-adapting Local-eddy diffusivity model within a variational multi-scale framework (VMS-WALE)[1, 2] and the QR-model [3]. The results have been compared to Direct Numerical Simulations (DNS) and after that, LES models have been used to study the stall phenomenon. Thus, the objective of this work is two-fold: i) to assess the performance of these LES models for predicting the flow with massive separation and, ii) to study the flow behaviour at near stall and stall angles in order to shed some light into the physics of this complex flow. To do this, we have performed LES of the flow past a NACA0012 at Reynolds number of $Re = 5 \times 10^4$ and at angles of attack (AOA) up to $12°$.

2 Mathematical and Numerical Model

The methodology for solving the flow past bluff bodies which exhibits massive separation has been described in detail in [4, 5]. The governing equations have been discretised on a collocated unstructured grid arrangement by means of second-order spectre-consistent schemes [6]. Such schemes are conservative, i.e. they preserve the symmetry properties of the continuous differential operators on any grid and even at high Reynolds numbers. For the temporal discretisation of the momentum equation a two-step linear explicit scheme on a fractional-step method has been used for the convective and diffusive terms, while for the pressure gradient term an implicit first-order scheme has been used. By filtering the Navier-Stokes system of differential equations, the subgrid (SGS) stress appears in the momentum equations which must be modelled in order to close the system. LES studies have been performed using two SGS models suitable for unstructured formulations: the QR-model [3] based on the invariants of the rate-of-strain tensor and the wall-adapting local-eddy viscosity model within a variational multi-scale formulation (VMS-WALE) [1, 2].

3 Results

Numerical simulations of the flow around a NACA0012 at Reynolds numbers $Re = U_{ref}C/\nu = 5 \times 10^4$ (defined in terms of the free-stream velocity U_{ref}, the airfoil chord C and the flow kinematic viscosity ν) and at $AOA = 8°$, $8.75°$, $9°$, $9.25°$ and $12°$ have been carried out. Solutions are obtained in a computational domain of dimensions $40C \times 40C \times 0.2C$ with the leading edge of the airfoil placed at $(0,0,0)$. The boundary conditions at the inflow consist of a uniform velocity profile (u,v,w)=($U_{ref}\cos AOA$, $U_{ref}\sin AOA$,0). At the outflow boundary, a pressure-based condition is imposed. No-slip conditions on the airfoil surface are prescribed. Periodic boundary conditions are used in the span-wise direction. For the assessment of the two SGS models, solutions have been obtained on two grids of 46417×16 planes ($\approx$0.74 million control volumes (MCVs)) and 149355×32 planes ($\approx$4.78MCVs), whereas DNS has been computed on a grid of 280876 × 96 planes (≈27MCVs). Although not shown here, this mesh has been analysed in order to test if it was capable of solving the smallest scales of the flow. Moreover, LES results at AOA from 8.75° to 12° have been computed in a set of two grids designed so as the refined zone in the suction side is well covered at all AOAs.

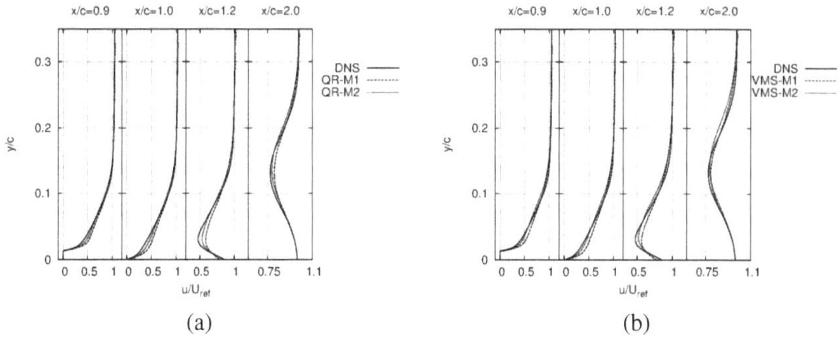

Fig. 1 LES assessment against DNS results at AOA=8°. Velocity profiles for (a) QR- and (b) VMS-WALE models.

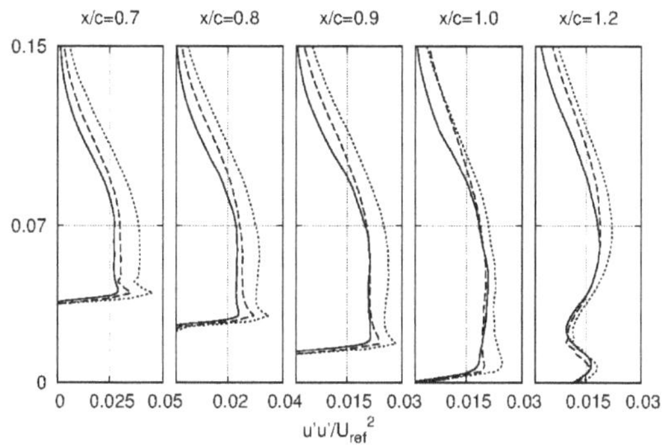

Fig. 2 Stream-wise velocity fluctuations for QR- and VMS-WALE models

3.1 LES Assessment

Although results were compared to a larger number of locations, for the sake of brevity we are only presenting results for four of them and for stream-wise velocity and its fluctuations (see figs. 1 and 2). Although for both models a very good prediction of stream-wise velocity is achieved, it is observed that VMS-WALE model presents a better trend as the mesh is refined and specially in the detached zones. Stream-wise velocity fluctuations are also shown in the figure. Contrary to first-order statistics which are well predicted even with the coarse grid, the finest grid is required for a better prediction of the velocity fluctuations. In general, VMS-WALE behaves better and its trend to predict DNS results as the grid is refined. Yet, it might be said that both models are good choices for predicting the flow with massive separation which is specially true for the finest grid.

3.2 LES Results at Stall AOAs

In order to gain insight into the coherent structures developed in the separated zone, these structures have been identified by means of the second invariant of the velocity gradient (Q-criterion) proposed by Hunt et al. [7]. These structures are shown in figure 3 at $AOA = 8.75^\circ, 9^\circ, 9.25^\circ$ and 12°. The flow separates laminarly from the airfoil surface near the leading edge and vortex breakdown occurs at the end of the laminar shear-layer as a consequence of the instabilities developed by the action of a Kelvin-Helmholtz mechanism. This mechanism of transition is similar to that observed in other bluff bodies such as the flow past a circular cylinder [5] or the flow past a sphere [4]. However, as the AOA increases, the separated region also increases and at the largest AOA, the flow is fully detached from the airfoil surface and the shedding of vortices is clearly observed.

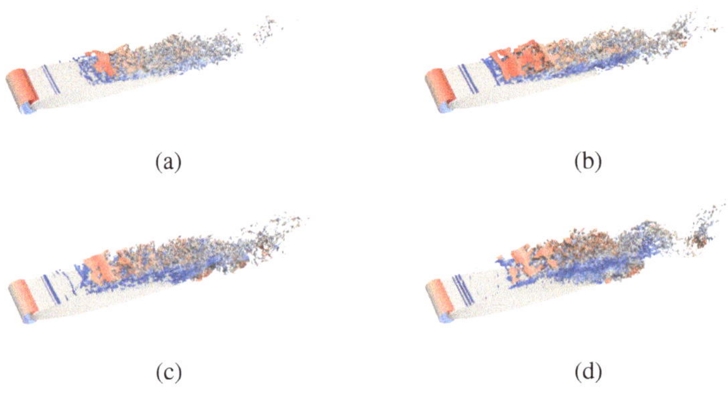

Fig. 3 Visualization of the flow structures by means of Q-isocontours ($Q = 30$) [7] coloured by the velocity magnitude (a) $AOA = 8.75^\circ$, (b) $AOA = 9^\circ$, (c) $AOA = 9.25^\circ$ and (d) $AOA = 12^\circ$

Figure 4 shows the transient evolution of the lift coefficient for the different AOAs. As can be observed at AOA=8.75° it oscillates with a very low fluctuation. This AOA corresponds with the highest value in the lift coefficient which is $C_l = 0.795$. When the AOA is slightly increased to AOA=9°, low-frequency fluctuations are observed. This sudden decrease of the lift coefficient is produced by the flow detachment from the airfoil surface. For larger AOAs, the flow oscillation occurs but at a higher frequency corresponding with the complete detachment of the flow and the shedding of vortices. The pattern observed in the lift behaviour at stall angles (AOA=9° and AOA=9.25°) is quite similar to that observed experimentally by Tanaka [8] at $Re = 1.3 \times 10^5$. In his experimental observations the change

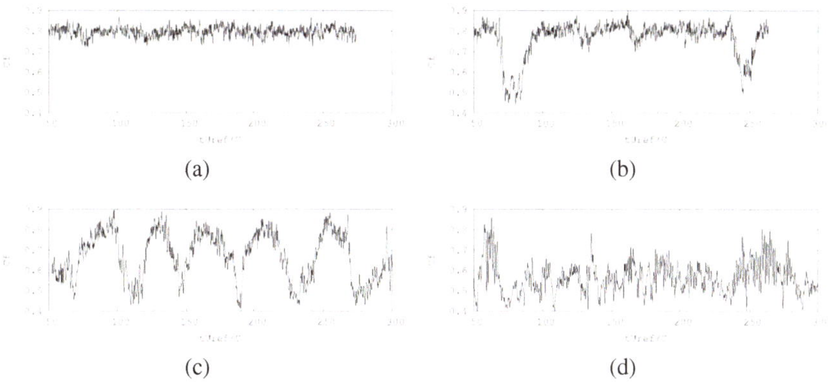

Fig. 4 Transient evolution of the lift coefficient. (a) 8.75°, (b) 9°, (c) 9.25° and (d) 12°.

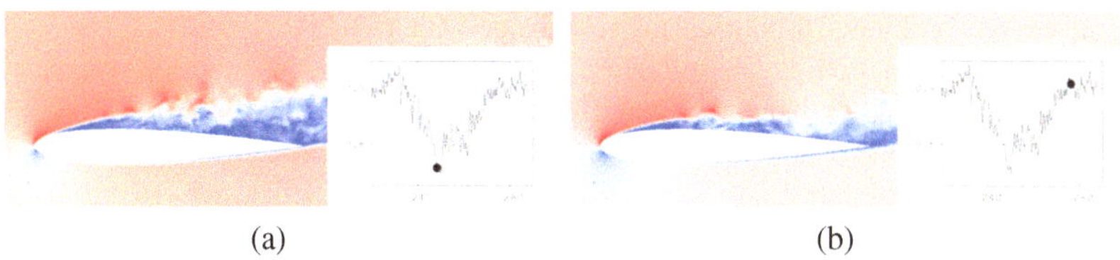

Fig. 5 Transient evolution of the flow at AOA=9°, from flow fully separated and minimum lift (a) to flow reattachment (b)

of behaviour from low-frequency oscillation to a higher frequency oscillation also occurred within a small variation in the AOA (0.2°), but at a larger AOA due to the larger Reynolds number.

This process (at AOA=9°) can be closer inspected in figure 5, where instantaneous velocity profiles when the flow is completely detached (see Figure 5a) and when the recirculation bubble is at its minimum (see Figure 5b) are plotted. In the figure, the instantaneous lift coefficient (see figure inset) is also shown.

It should be stressed that the non-dimensional frequency measured for this process at the beginning of the stall is $St = f\sin(AOA)C/U_{ref} = 0.004$ (here $f = 0.0247\,s^{-1}$) with a $C_{l,rms} = 0.115$. This corresponds with a thin airfoil- trailing edge type stall [9].

4 Conclusions

The assessment of two subgrid-scale models for predicting the flow past an airfoil with flow separation have been performed. Both models (VMS-WALE and QR-model) perform quite well being capable of predicting separation and transition to turbulence, as well as the fully developed turbulence in the wake, although VMS-WALE model performs slightly better than QR-model. The study of the flow at near stall and in stall angles has shown that at the beginning of the stall (AOA=9°) the flow behaves very unsteady with a low-frequency oscillation which has been measured, being of St = 0.004 and with a fluctuation in the lift coefficient $C_{l,rms} = 0.115$ which corresponds with a thin airfoil-trailing edge type stall. Results agree quite well with literature observations of this phenomena for airfoils exhibiting thin airfoil-trailing edge stall.

Acknowledgements. This work has been financially supported by the "Ministerio de Economía y Competitividad, Secretaría de Estado de Investigación, Desarrollo e Innovación", Spain (ref. ENE2009-07689 and ENE2010-17801) and by the Collaboration Project between Universidad Politècnica de Catalunya and Termo Fluids S.L.

References

1. Nicoud, F., Ducros, F.: Subgrid-scale stress modelling based on the square of the velocity gradient tensor. Flow, Turbulence and Combustion 62, 183–200 (1999)
2. Hughes, T.J.R., Mazzei, L., Jansen, K.E.: Large eddy simulation and the variational multiscale method. Computing and Visualization in Science 3, 47–59 (2000)
3. Verstappen, R.: When does eddy viscosity damp subfilter scales sufficiently? In: Quality and Reliability of Large-Eddy Simulations II. ERCOFTAC Series, vol. 16, pp. 421–430 (2011)
4. Rodríguez, I., Borrell, R., Lehmkuhl, O., Pérez-Segarra, C.D., Oliva, A.: Direct Numerical Simulation of the Flow Over a Sphere at Re = 3700. Journal of Fluid Mechanics 679, 263–287 (2011)
5. Lehmkuhl, O., Rodríguez, I., Borrell, R., Oliva, A.: Low-frequency unsteadiness in the vortex formation region of a circular cylinder. Physics of Fluids 25, 085109 (2013)
6. Verstappen, R.W.C.P., Veldman, A.E.P.: Symmetry-Preserving Discretization of Turbulent Flow. Journal of Computational Physics 187, 343–368 (2003)
7. Hunt, J.C.R., Wray, A.A., Moin, P.: Eddies, stream and convergence zones in turbulent flows. Technical Report CTR-S88, Center for turbulent research (1988)
8. Tanaka, H.: Flow visualization and PIV measurements of laminar separation bubble oscillating at low frequency on an airfoil near stall. In: 24th International Congress of the Aeronautical Sciences, pp. 1–15 (2004)
9. Broeren, A.P., Bragg, M.B.: Low-frequency flowfield unsteadiness during airfoil stall and the influence of stall type. In: AIAA Paper, AIAA–98–2517 (1998)

Instability of the Helical Tip Vortices behind a Single Wind Turbine

Sasan Sarmast*, Philipp Schlatter, Stefan Ivanell, Robert F. Mikkelsen, and Dan S. Henningson

Abstract. A numerical study on a single wind turbine wake has been carried out focusing on the instability properties of the trailing tip vortices shed from the turbine blades. The numerical model is based on large-eddy simulations (LES) of the Navier–Stokes equations together with the actuator line method to simulate the wake behind the Tjæreborg wind turbine. The wake is perturbed by low amplitude stochastic excitations located in the neighborhood of the tip spiral, giving rise to spatially developing instabilities. Dynamic mode decomposition (DMD) is then utilized for identification of the coherent flow structures. The DMD results indicate that the amplification of specific waves along the spiral is responsible for triggering the instability leading to wake breakdown. Two types of dynamic structures dominates the flow; low and high frequency groups. Examination of these structures reveals that the dominant modes have the largest spatial growth.

1 Introduction

Wind turbines are nowadays clustered inside wind farms. Depending on wind direction, turbines are fully or partially influenced by the upstream turbine wakes. An unwanted but inevitable effect, however, is that the efficiency of the interior turbines decreases due to the velocity deficit and the turbulence intensity increases due to the interaction from the wakes of the surrounding wind turbines. As a consequence, dynamic loadings are increased which may excite the resonance frequency in the

Sasan Sarmast · Philipp Schlatter · Stefan Ivanell · Dan S. Henningson
Swedish e-science Research Centre (SeRC), Linné FLOW Centre, KTH Mechanics, Stockholm, Sweden
e-mail: sarmast@mech.kth.se

Stefan Ivanell
Gotland University (HGO), Visby, Sweden

Robert F. Mikkelsen
DTU wind energy, Lyngby, Denmark

* Corresponding author.

M. Hölling et al. (eds.), *Wind Energy – Impact of Turbulence*,
Research Topics in Wind Energy 2,
DOI: 10.1007/978-3-642-54696-9_25,

structural parts of the individual wind turbines and increase the fatigue loads. Therefore, the detailed knowledge of the unsteady wake development helps to understand problems related to performance, vibrations, noise and structural load. These issues are studied by researchers, and their common goal is cost reduction through optimizing the efficiency of wind farms and minimizing the loads.

The wake behind a wind turbine consists of a strong root and distinct tip vortices. The tip/root vortex system is usually unstable, and due to self-induced instability mechanisms and ambient atmospheric turbulence, it eventually breaks down. Early study of the helical vortices dates back to the work by Widnall [10] where she provided the analytical framework to study the linear stability of the helical vortices for the inviscid flows. Widnall proves the existence of at least three different instability mechanisms. Using a single twisted helix, she found that the vortex dynamics are unstable due to long– and short–wave instability mechanisms, as well as to a mutual inductance. In an experimental attempt, Felli et al. [1] made extensive investigations on propeller wakes. They acknowledged the traces of all the three types of instability mechanisms in their experiments. Walther et al. [9] performed series of direct numerical simulations on the stability of the helical vortices. They confirmed that the same instability mechanisms observed by Widnall can also be found in the presence of viscosity. Recent numerical stability analysis by Leishman et al. [3] and Ivanell et al. [2] shows that the instability is dispersive and they reported that the spatial growth arises only for some specific frequencies and type of modes with specific wave-numbers equal to half-integer multiples of the number of blades, *i.e.* for a three blade configuration occurring at $1\frac{1}{2}, 3\frac{1}{2}, 7\frac{1}{2}, \cdots$. Ivanell reported that the largest spatial growth rate occurs where the oscillation of neighboring vortex spirals are out-of-phase and the mutual inductance instability is associated with the vortex pairing.

The present study focuses on the stability properties of the tip vortices and the mechanisms leading to wake breakdown. The basic wake behind an horizontal axis wind turbine is computed by combination of the unsteady incompressible Navier–Stokes solver (EllipSys3D) and the actuator line (ACL) method. Dynamic mode decomposition (DMD) technique [6, 7] is then utilized for identification of the coherent structures in the near wake.

2 Problem Setup

The actuator line (ACL) method [8] is a fully three-dimensional and unsteady aerodynamic model for simulating the wind turbine wakes. The flow field around the rotor is governed by the full three-dimensional incompressible Navier–Stokes equations, while the influence of the rotating blades is accounted using body force. The force is determined using the blade element method (BEM) combined with tabulated airfoil data. The method is implemented into the EllipSys3D code [4, 5].

The computational domain are discretized in an axisymmetric polar grid with 12.6 million grid points. In the vicinity of the actuator lines, referred to as the near domain, approximately 9 million grid points are used in order to capture the gradients and fully resolve the near wake dynamics. The simulation is performed in a rotating frame of reference where the actuator lines are standing still. Considering the boundary conditions, the velocity at the inlet is assumed to be uniform. A convective outflow boundary condition is used at the outlet. On the lateral walls, the periodic boundary conditions are imposed. Figure 1 shows the computational domain where the outer cylindrical walls are chosen far from the rotor to assure very low blockage effect. The current simulations are performed based on the Tjæreborg wind turbine operating at the optimum power condition ($C_p = 0.49$) and a wind speed of $10\,m/s$ corresponding to tip speed ratio of 7.07. Note that all the simulation parameters are normalized by the wind speed $U_\infty = 10\,m/s$ and the blade length $R = 30.5\,m$.

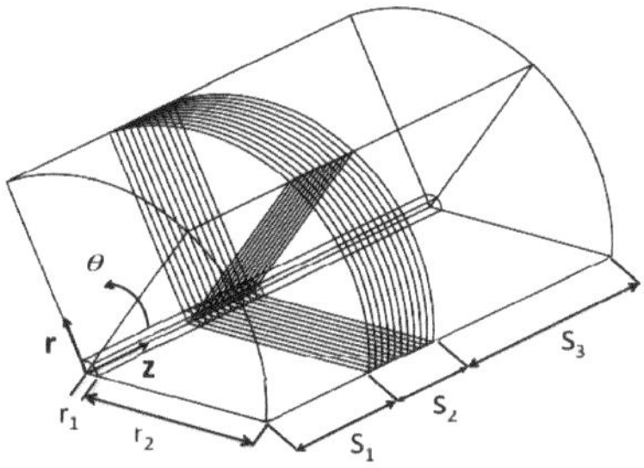

parameter	Length (in R)
r_1*	1.4
r_2	13.6
S_1	12.5
S_2*	8.5
S_3	22

* Near domain

Fig. 1 The figure shows schematically the multi-block mesh topology (48 blocks). r_1 and S_2 represent the near domain where the grid is distributed equidistantly in all the directions. In r_2, S_1 and S_3 regions, the grids are stretched towards the outer boundaries in both axial and radial directions. The actuator line is discretized by 70 grid points to resolve the blade accurately. The flow direction corresponds to the positive z-axis and the rotor is located inside near domain at $z = 14R$.

To proceed with the simulation, first, the steady base flow is obtained (Figure 2(a)) with the simulation time $t = t_{ph}\frac{U_\infty}{R} = 225$. Then, additional perturbations are introduced at the downstream of the blade tips. The additional perturbations trigger the instability mechanisms which results in the wake breakdown (Figure 2(b)). When the flow reaches its fully developed instationary condition $t = 300$, we start to collect the snapshots.

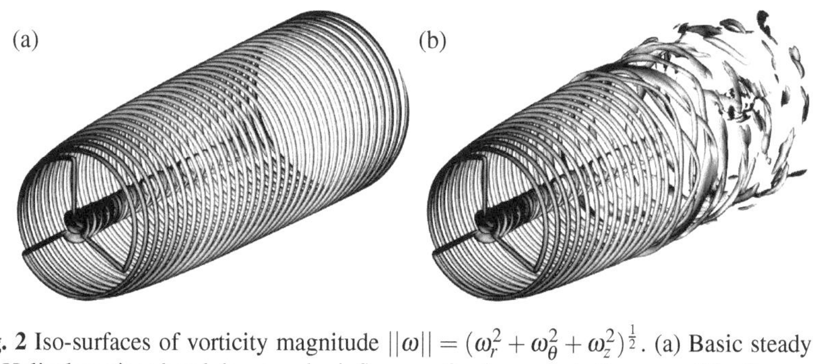

Fig. 2 Iso-surfaces of vorticity magnitude $||\omega|| = (\omega_r^2 + \omega_\theta^2 + \omega_z^2)^{\frac{1}{2}}$. (a) Basic steady state. (b) Helical vortices breakdown under influence of noise perturbations amplification.

A controlled way of disturbing the helical vortices is introducing low amplitude perturbation to the steady base flow. The spatial shape of the body force is assumed to be a Gaussian distribution where its temporal part is considered as a low-pass filtered harmonic excitation. The force is applied only in the axial direction and the amplitude is tuned to obtain the perturbation fields with turbulence intensity $I = 0.1\%$ near the rotor tip. The applied perturbation contains noise with a uniform energy distribution over all frequencies lower than the cutoff frequency. The cutoff frequency is chosen large enough to remove the physically damped higher frequencies in the flow.

The dissection of the flow field into subdomain allows us to concentrate on the localized phenomena. The length of the subdomains and their positions are important parameters to study two phenomena in the wake: the mechanism of instability onset, and the nonlinear effects on the wake breakdown. The former study introduces the linear subdomain and the choice of nonlinear subdomain is based on where the tip vortices start to deform from the helical path and start to interact mutually. The linear and nonlinear subdomains with their toroidal shapes are depicted in Figure 3.

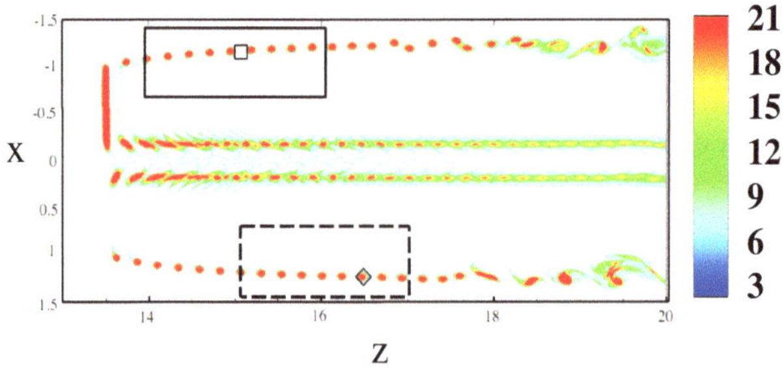

Fig. 3 Contours of vorticity magnitude in a x/z-plane cut show the development of tip spiral instability. The solid and dashed line indicate the position of the linear and nonlinear subdomains; while, the small square and diamond define the location of the extracted probes in the linear and nonlinear subdomains, respectively.

3 Dynamic Mode Decomposition

The dynamic mode decomposition (DMD) is used to describe the underlying mechanism of the wake breakdown. DMD splits the flow into different modes where each one of them oscillates at a certain frequency and growth rate. Given a sequence of snapshots (flow fields) x_k collected with an equidistant time-interval Δt, we assume a constant linear mapping $\tilde{A}$ between each of the snapshots, $x_{j+1} = \tilde{A}x_j$. Considering the entire snapshot sequence $X_0^n = [x_0, x_1, \ldots, x_n]$, we can write

$$\tilde{A}X_0^n = X_1^{n+1}. \tag{1}$$

For a sufficiently large number of snapshots, we can assume that the snapshots are no longer linearly independent, so we can write the last snapshot as a linear combination of the previous ones. Thus, we have

$$\tilde{A}X_0^n = X_1^{n+1} = X_0^n C + \varepsilon e^T, \tag{2}$$

where the C is a companion type matrix and it contains the linear combination coefficients. The symbol ^{T} represents conjugate transpose, $e = [0,\ldots,0,1]^T$ and ε contains the dynamic residuals. In fact, we project the matrix $\tilde{A}$ over the snapshot sequence. To improve the robustness of the algorithm, Schmid [7] made an improvement by projecting the system onto the proper orthogonal decomposition (POD) basis. To proceed, we employ an economy size singular value decomposition (SVD) of the snapshots matrix $X_0^n = U\Sigma W^T$, where the columns of U and W are the left and right singular vectors, respectively. The diagonal entries of Σ, is known as the singular values. We can then rewrite equation 2 as

$$\tilde{A}U \approx U\tilde{C}, \tag{3}$$

where $\tilde{C}$ is the transformed companion matrix and it can be computed from $\tilde{C} = U^T X_0^n W \Sigma^{-1}$. Using the spectral expansion, we can find the eigenvalues Λ and eigenvectors Y of the transformed companion matrix, $\tilde{C} = Y\Lambda Y^{-1}$. Now, we can approximate the eigenvalues and eigenvectors of the linear map $\tilde{A}$ as Λ and $\Phi = UY$, respectively.

In DMD procedure, we would like to find an expansion of the snapshot sequence of the form

$$x_k = \sum_{j=1}^{n} \lambda_j^k \phi_j d_j, \quad k = 0,1,\ldots,n-1, \qquad \text{or} \qquad X_0^n = VDS, \tag{4}$$

where the matrix $V = UY$ contains the dynamic modes, D is a diagonal matrix and it is related to the amplitudes of the DMD modes and S is the so-called Vandermonde matrix and it is defined as

$$S = \begin{bmatrix} 1 & \lambda_1 & \lambda_1^2 & \ldots \\ 1 & \lambda_2 & \lambda_2^2 & \ldots \\ 1 & \lambda_3 & \lambda_3^2 & \ldots \\ \vdots & & \vdots & \ddots \end{bmatrix}, \tag{5}$$

where λ_j are the eigenvalues of the matrix $\tilde{C}$. To obtain the matrix D, without inverting the Vandermonde matrix S, we substitute the singular value decomposition of the snapshots into equation 4,

$$D^{-1} = SW\Sigma^{-1}Y. \tag{6}$$

Since D is a diagonal matrix, it can be obtained by inverting the diagonal entries of D^{-1}. Finally, the DMD amplitudes Φ can be obtained from the diagonal components of $|D|^2$, if the eigenvectors of the transformed companion matrix, Y, are normalized to unity.

3.1 Convergence Analysis and Choice of the Data-Set

Important parameters of the dynamic mode decomposition are the time interval between two consecutive snapshots Δt and the number of snapshots n. The time interval Δt should be adjusted considering the Nyquist criterion. According to this criterion, the maximum frequency that we can resolve is $f = \frac{1}{2\Delta t}$. In fact, this frequency depends on the highest physically-relevant frequency in the flow. The time interval $\Delta t = 0.025$ is chosen which resolves a maximum non-dimensional frequency corresponding to $St = fR/V_\infty$ =20. On the other hand, the number of snapshots has to be chosen such that, the snapshots become linearly dependent. This is tested using the dynamic residuals obtained from the least square problem (Equation 2). Figure 4 shows the residuals of the linear and nonlinear subdomains: The overall trend shows a reduction of the residual values as the number of snapshots increases. Even though, 800 number of snapshots gives us the converged statistics, the following results are based on 1000 number of snapshots with the time interval $\Delta t = 0.025$.

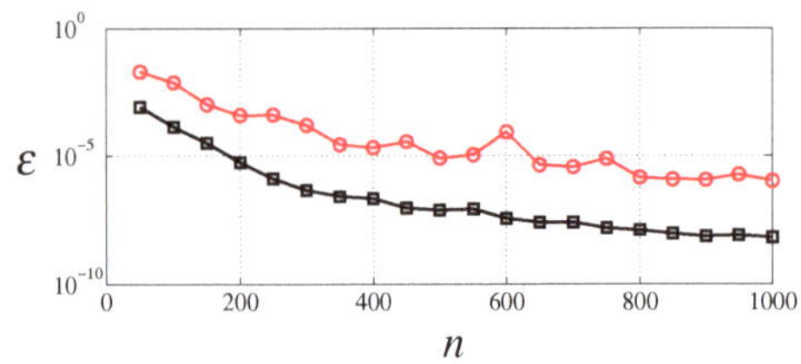

Fig. 4 Residual norm of the dynamic mode decomposition. Red circles and black squares represent the dynamic residuals of the linear and nonlinear subdomains, respectively.

4 Results

The results are based on the DMD algorithm presented in the section 3. DMD is applied on the snapshot sequences obtained from the linear and nonlinear subdomains. Figure 5 shows the eigenvalues of the transformed companion matrix $\tilde{C}$, where all lie approximately on the unit circle $|\lambda_j| \approx 1$. This implies that the flow is in equilibrium and no significant temporal growth or decay is present. An eigenvalue at (1,0) indicated by the blue symbol corresponds to a steady mean flow (temporally averaged flow field of the data sequence). The coloring (red to white) of the eigenvalues is related to the dynamic modes amplitudes (Φ) and it is intended to separate dynamic structures (red) from weaker ones (white). The modes with higher amplitudes determine the presence of specific dynamics in the underlying flow physics.

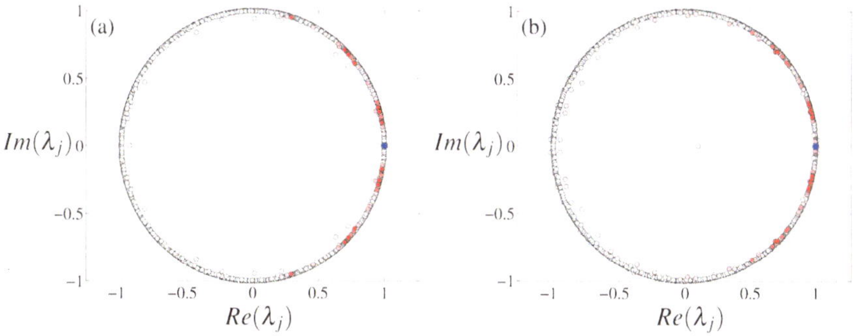

Fig. 5 DMD eigenvalues of (a) linear and (b) nonlinear subdomains. The colors vary smoothly depending on the DMD amplitudes from red (high) to white (low). The blue symbol positioned at $(1,0)$ depicts the mean flow eigenvalue.

4.1 Power Spectra and Temporal Dynamic Structures (Modes)

Traditionally, the temporal information of the wake is obtained by using the time signal probe; for instance, such a probe measures the instantaneous velocity components at specific locations. The analysis of the time signal helps us to understand the flow dynamics. In this study, we gather the information of two local probes positioned at linear subdomain (Figure 6 (a)) and nonlinear subdomain (Figure 6 (b)). The spectral information of the probes together with their corresponding DMD spectrums are depicted in Figure 6 (c,d). In DMD spectrums, the amplitude of the dynamic modes are plotted as a function of the normalized frequency using the convention $\omega_j = \mathrm{Im}(\log(\lambda_j)/\Delta t)$. There is a general agreement between both spectrums. However, we can observe that the distinct single peaks are not exactly matching, instead the cluster of peaks overlap. This can be explained through the limitation of a number of discrete snapshots, and the specific location of the probes in the wake. Two pronounced clusters of peaks dominate the flow in the linear and nonlinear subdomains, see Figure 6 (c-d), with frequencies around $St = 1.6$ and $St = 5$. From now on, the former group is called the low frequency group and the latter one as high frequency group. In Figure 6 (d), it is clear that the distribution of low and high frequency groups is broadened and more modes are triggered in the nonlinear subdomain. The presence of higher frequency modes is related to the fact that all linear combinations of frequencies excite higher modes, for instance, the group of frequency around $St = 7$ can be the linear interaction of low and high frequency groups.

To better understand the dynamics of the flow, the leading dynamic structures of the linear subdomain are shown in Figure 7. In low frequency group, each helix is divided into three lobes corresponding to wave number of $K \approx 3/2$. Similarly, the high frequency group modes are constituted of nine lobes corresponding to wave number of $K \approx 9/2$. There are also some less energetic modes which do not fall within the above categories, but we will not discuss them further.

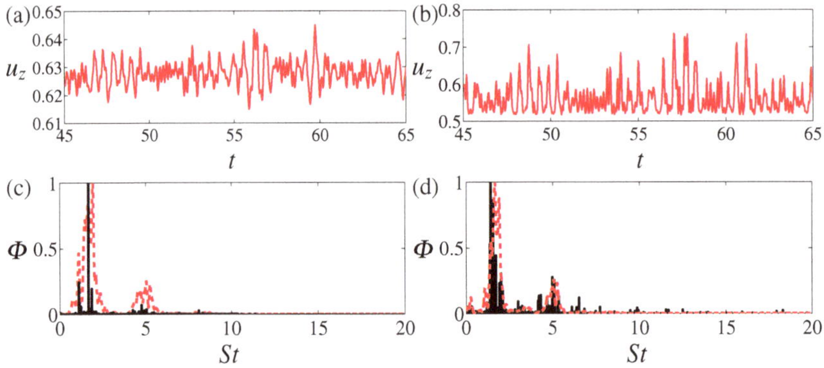

Fig. 6 (a,b) Time signal probes located at the linear subdomain ($[r, \theta, z] = [1.23, 1.57, 15.0]$) and nonlinear subdomain ($[r, \theta, z] = [1.23, 1.57, 16.5]$). (c,d) Amplitude distribution of the dynamic modes together with the power spectrum of the time signal probe in the linear and nonlinear subdomains, respectively. Black bars represent the DMD spectrum and the red dashed line are the power spectrum of the time signal probes.

Fig. 7 (a) Energetic low frequency group structures (modes) and (b) energetic high frequency group structures of the linear subdomain

4.2 Spatial Analysis Based on Temporal Modes

A closer look at the velocity components of the dominant dynamic modes indicates the presence of a wave-like motion, propagating along the spirals. In particular, if we follow the maximum amplitude along the spiral, we can observe a spatial exponential amplification; the maximum amplitude is defined as

$$w(z) = \max_{r,\theta} \sqrt{u_r^2 + u_\theta^2 + u_z^2} \qquad r, \theta, z \in \text{a spiral trajectory,} \tag{7}$$

where u_r, u_θ and u_z are the radial, azimuthal and axial velocities of the modes. The spatial growth rate (denoted as α) is defined as the slope of the maximum amplitude curve in the logarithmic scale $\alpha = \frac{d(log(\mathbf{w}))}{dz}$. Examining the first 400-energetic modes and obtaining their individual growth rates show clustering of the growth rate within a sinusoidal type of variation (Figure 8), where the most energetic temporal

DMD modes are also the most spatially unstable modes. The result shows more scattered behavior as less and less energetic modes are considered. The result is in good agreement with the previous observations of Leishman et al. [3], and Ivanell et al. [2] where the wave numbers corresponding to maximum growth of instability occurs at half-integer multiples of the number of blades; similarly, the minimum instability growth happens at wave numbers of multiple of blades number.

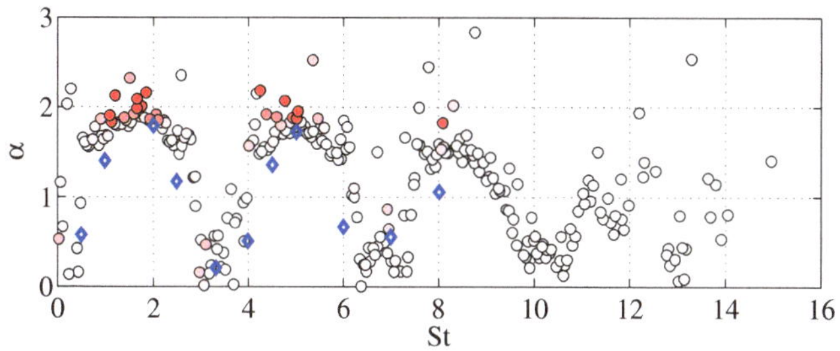

Fig. 8 Spatial growth rate as function of Strouhal number. Circles correspond to leading DMD modes in the linear subdomain. The colors vary smoothly depending on the DMD amplitudes from red (high) to white (low). The blue diamonds are the result of stability analysis performed by Ivanell et al. [2].

Based on the modal expansion of snapshots presented in Equation (4), it is possible to introduce a filtering technique that removes presumably less important structures from the flow, thus yielding a low-order model of the wind turbine wake. A superposition of sufficiently many energetic modes would capture the dynamics of large-scale dynamics and could reproduce the main features of the original snapshot sequence. Figure 9 demonstrates the filtering approach, where the first 50 energetic modes are included to reconstruct the original flow field. Two probes are extracted from the reconstructed flow fields in linear and nonlinear subdomains. It is observed that the reconstructed flow fields still capture the dynamic of the large scales, while it filters out the higher frequencies from the system in a global way.

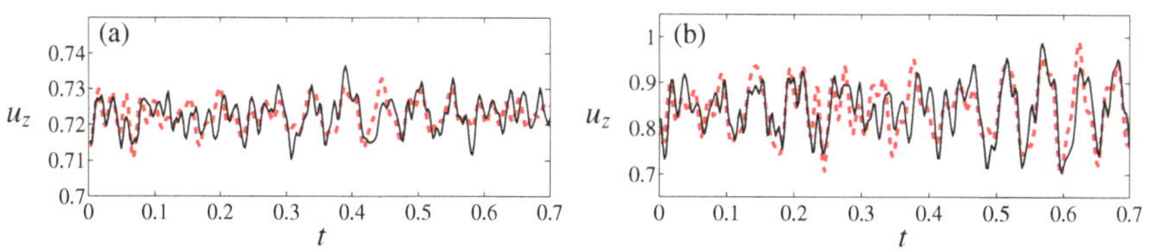

Fig. 9 Time evolution recorded in two positions at (a) linear and (b) non-linear subdomains. The position of the probes are shown in Figure 3. The black line represents the original signal and the dashed red line is the reconstructed signal from 50 dominant DMD modes.

5 Conclusions

To determine the stability properties of wind turbine wakes a numerical study on the stability of the tip vortices behind the Tjæreborg wind turbine has been carried out. The numerical model is based on large-eddy simulations of the Navier–Stokes equations using the actuator line method to generate the wake. The flow is then perturbed by inserting stochastic excitation leading to wake breakdown. By analyzing the temporal flow fields using the dynamic mode decomposition (DMD) approach, we found that the amplification of specific waves along the spiral is responsible for triggering the instability leading to wake breakdown. These dominant structures can be categorized by their modal shapes and frequencies. Two group of modes are then identified in the wake, the low frequency and high frequency groups. Examining these modes, it is shown that the high energetic temporal modes have the highest spatial growth. Furthermore, a filtering procedure based on DMD method is demonstrated.

References

1. Felli, M., Camussi, R., Di Felice, F.: Mechanisms of evolution of the propeller wake in the transition and far fields. J. Fluid Mech. 1(1), 1–49 (2011)
2. Ivanell, S., Mikkelsen, R., Sørensen, J.N., Henningson, D.S.: Stability analysis of the tip vortices of a wind turbine. Wind Energy 13(8), 705–715 (2010)
3. Leishman, G., Bhagwat, M.J., Ananthan, S.: The vortex ring state as a spatially and temporally developing wake instability. J. Am. Helicopter Soc. 49 (2004)
4. Michelsen, J.A.: Basis3d - a platform for development of multiblock pde solvers. Technical Report AFM 92-06, Dept. of Fluid Mechanics, Technical University of Denmark, DTU (1992)
5. Sørensen, N.N.: General perpose flow solver applied to flow over hills. PhD thesis, Risø National Laboratory, Roskilde (1995)
6. Rowley, C.W., Mezić, I., Bagheri, S., Schlatter, P., Henningson, D.S.: Spectral analysis of nonlinear flows. J. Fluid Mech. 641(1), 115–127 (2009)
7. Schmid, P.J.: Dynamic mode decomposition of numerical and experimental data. J. Fluid Mech. 656(1), 5–28 (2010)
8. Sørensen, J.N., Shen, W.Z.: Numerical modeling of wind turbine wakes. J. Fluid Eng. 124 (2002)
9. Walther, J.H., Guenot, M., Machefaux, E., Rasmussen, J.T., Chatelain, P., Okulov, V.L., Sørensen, J.N., Bergdorf, M., Koumoutsakos, P.: A numerical study of the stabilitiy of helical vortices using vortex methods. Journal of Physics: Conference Series 75, 012034 (2007)
10. Widnall, S.E.: The stability of a helical vortex filament. J. Fluid Mech 54(4), 641–663 (1972)

High-Resolution Offshore Wake Simulations with the LES Model PALM

Björn Witha, Gerald Steinfeld, and Detlev Heinemann

Abstract. Large-eddy simulations of wind turbine wakes have been performed with the LES model PALM. A comparison of the implemented actuator disk and actuator line models shows differences in the near wake which diminish in the far wake several diameters downstream of the rotor plane. Furthermore, the results of wake simulations with different inflow boundary conditions are presented. The major drawback of cyclic horizontal boundary conditions, which are often used for wake simulations, is that the inflow used will be the one already modified by the wind turbines. With non-cyclic boundary conditions and a laminar inflow, the inflow remains undisturbed and constant in time. However, a very large model domain is required so that a turbulent flow can develop. The application of a turbulence recycling method results in a well developed turbulent boundary layer already at the inflow so that the model domain can be significantly reduced.

1 Introduction

The wake of a wind turbine is characterized by a significant velocity deficit as the turbine extracts momentum from the mean flow. In wind farms this leads to a reduced power output for subsequent downstream turbines. Furthermore, the wakes feature a significantly enhanced turbulence intensity which results in an increased load for downstream turbines. Thus, wake modeling is a key task for the energy yield prediction of wind farms as well as for the design of new wind farms.

Mostly either simple engineering wake models [2] or Reynolds-Averaged Navier-Stokes Simulations (RANS [6]) are used for wake calculations. In these models, the turbulence is not resolved but completely parameterized. Current powerful High-Performance Computing (HPC) clusters allow for turbulence resolving Large-Eddy Simulations (LES) of wind turbine wakes and intra wind farm wakes. With LES it is possible to investigate the interaction between the atmospheric boundary layer and

Björn Witha · Gerald Steinfeld · Detlev Heinemann
ForWind, Carl von Ossietzky Universität Oldenburg,
Ammerländer Heerstraße 136, 26129 Oldenburg, Germany
e-mail: bjoern.witha@forwind.de

M. Hölling et al. (eds.), *Wind Energy – Impact of Turbulence*,
Research Topics in Wind Energy 2,
DOI: 10.1007/978-3-642-54696-9_26,

the wind turbine wakes as well as the interaction between the individual wakes in a wind farm. An increased knowledge about wake flows can be expected, which can be applied to develop better parameterizations of wake flows in RANS models or in engineering models for a more reliable energy yield prediction. However, in LES the impact of the wind turbines on the atmospheric flow still has to be parameterized. Two wind turbine models, an actuator disk and an actuator line model have been implemented in the LES model PALM. They will be discussed and compared in Section 2. Another challenge for LES of wake flows is the appropriate choice of inflow boundary conditions, which will be the topic of Section 3.

2 Wake Simulations with the LES Model PALM

In this study, simulations have been performed with the parallelized LES model PALM [7], which shows a very high performance and scalability on all state-of-the-art parallel architectures. Two wind turbine models have been implemented in PALM: a rather simple actuator disk model (ADM) and a much more complex Actuator Line Model (ALM). The ADM after [1] represents the impact of the rotor blades as an additional homogeneous momentum sink with a thrust force F acting against the mean flow:

$$F = -0.5\, C_T\, A \left(\frac{1}{1-a}\, u_r \right)^2 \tag{1}$$

with the thrust coefficient C_T, the rotor area A, the induction factor a and the temporally and rotor disk averaged velocity in direction of the mean flow u_r. The ADM with its uniformly loaded actuator disk represents a strong simplification and neglects rotational effects but is widely used for RANS and LES due to its simplicity and capability to deliver reasonable results with rather coarse grids [5].

The ALM after [8] describes the rotor blades as lines divided into segments. Lift and drag forces are calculated for each of the blade segments so that a non-uniformly distributed load and rotational effects are accounted for. The force F acting on a blade segment is:

$$\mathbf{F} = 0.5\, v_{rel}^2\, c\, dr\, (C_L\, \mathbf{e_L} + C_D\, \mathbf{e_D}) \tag{2}$$

with the local velocity relative to the rotating blade v_{rel}, the chord length c, the length of the rotor segment dr, the lift and drag coefficients C_L and C_D and the unit vectors in the directions of the lift and drag $\mathbf{e_L}$ and $\mathbf{e_D}$. The ALM is much more sophisticated and delivers more realistic results than the ADM [5], but requires significantly more computing time compared to the ADM. This is mainly due to the requirement of a much smaller time step to ensure that the rotor tip displacement is shorter than one grid cell length during one time step.

A comparison of the implemented ADM and ALM for the simulation of a single wake has been carried out for a turbine with 80 m hub height and 104 m rotor diameter. The model domain consists of 1536 x 256 x 64 grid points with a grid spacing of 4 m. As can be seen from Fig. 1, the wake characteristics in terms of the mean horizontal velocity are roughly similar for both methods. This holds especially

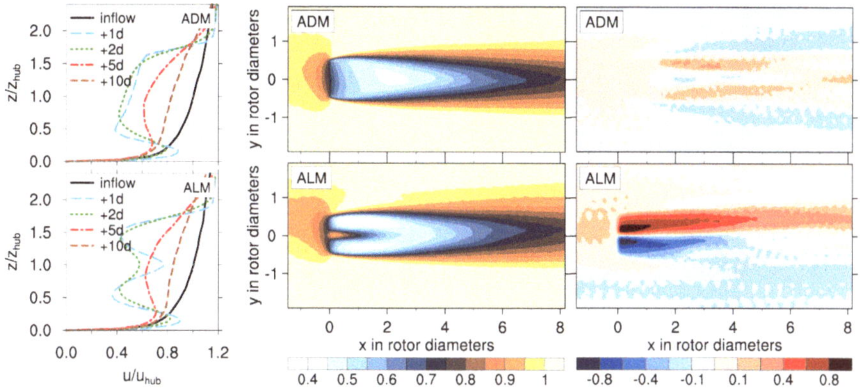

Fig. 1 Comparison of the single wake flow with ADM (top) and ALM (bottom). Left: time-averaged vertical profiles of the horizontal velocity u (normalized with the inflow velocity at hub height) upstream and at several locations downstream of the turbine, center: horizontal cross-section of the time-averaged and normalized horizontal velocity u in hub height; right: horizontal cross-section of the time-averaged vertical velocity w in hub height.

for the far wake from about 3 rotor diameters (D) downstream of the turbine. In the near wake the ALM shows a more detailed structure of the flow field than the ADM. Whereas the velocity deficit in the ADM simulation is rather uniform with a maximum in the center of the wake, the ALM simulation shows a circular maximum downstream of the outer parts of the rotor disk. Behind the center of the rotor disk, the wind speed is even higher than upstream of the turbine. However, it has to be taken into account that the turbine parameterizations do not include the effects of the tower and the nacelle. The ALM simulation features significant up- and downdrafts downstream of the outer parts of the rotor disk, reverse to the motion of the rotor blades. This is not visible in the ADM simulation, as the ADM does not consider rotation. The mean velocity profile in the far wake at a distance of 4 D downstream of the turbine or more is independent of the applied turbine parameterization.

3 Comparison of Inflow Boundary Conditions for Wake Simulations

To demonstrate the advantages and disadvantages of the various inflow conditions, a set of simulations has been carried out with the same turbine as in the previous paragraph. The model domain consists of 768 x 256 x 128 grid points with a grid spacing of 4 m. As the detailed wake structure is not important for this study, the less computationally intensive ADM is used.

Typically, cyclic (periodic) horizontal inflow boundary conditions are used in LES. For many applications, this is appropriate or even helpful, as turbulence can freely develop and is not effected by the side walls. When investigating wind turbine wakes and especially wind farm wakes, the periodic boundaries may lead to

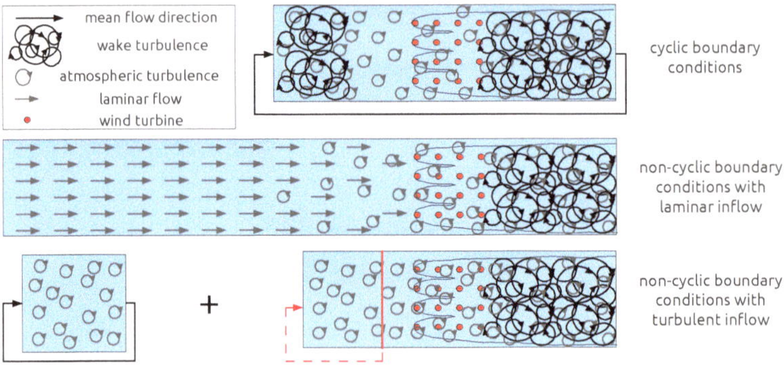

Fig. 2 Schematic of the different inflow boundary conditions in PALM

problems, as the turbulence generated in the wake of a wind farm would re-enter the model domain and modify the flow upstream of the wind farm, so the flow reaching the wind farm is no longer undisturbed ([3], see Fig. 2). In fact, an infinite row of turbines or wind farms is simulated. This is illustrated by the simulation results in Fig. 3a). A high level of (wake) turbulence is visible not only downstream but also upstream of the turbine.

With non-cyclic boundary conditions the wake turbulence is advected out of the model domain and the inflow remains undisturbed and constant in time. However, the inflow is laminar, as a constant wind speed is prescribed at the inflow, so the LES approach fails. This is clearly shown in Fig. 3b) with no fluctuations visible outside of the wake. The wake itself has a very smooth shape indicating the laminar flow as well. Several D downstream the wake breaks down into wave-like structures indicating the transition from laminar to turbulent flow. To ensure that the flow reaching the turbine is fully turbulent and quasi-stationary, a very long model domain is required (Fig. 2). Thus, the non-cyclic simulation with laminar inflow has been repeated with a ten times larger model domain in streamwise direction. To make this simulation feasible, a larger grid spacing of 10 m was used. As the results in Fig. 3c) show, the atmospheric turbulence is now well developed. In this case a fetch of about 15 km was necessary to reach a quasi-stationary turbulent boundary layer.

To avoid such a long model domain, a turbulent inflow has been realized in PALM which is based upon a recycling method after [4] (Fig. 2). At first, a precursor run with cyclic boundary conditions is performed. It does not contain any turbines and has a much smaller domain size compared to the main run (but large enough to resolve all scales of turbulence). To ensure a fully developed turbulent boundary layer, the simulation time of the precursor run has to be sufficiently long. The final results of the precursor run are then checked with respect to stationarity and turbulence statistics by looking at time series and spectra. The main run uses the data of the precursor run for initialization. The data is repeatedly mapped to the main run domain unless it is completely filled. The time- and domain-averaged profiles from the last time step of the precursor run are used as inflow profiles for the main run. In the main simulation the turbulence is constantly recycled which means that the turbulent

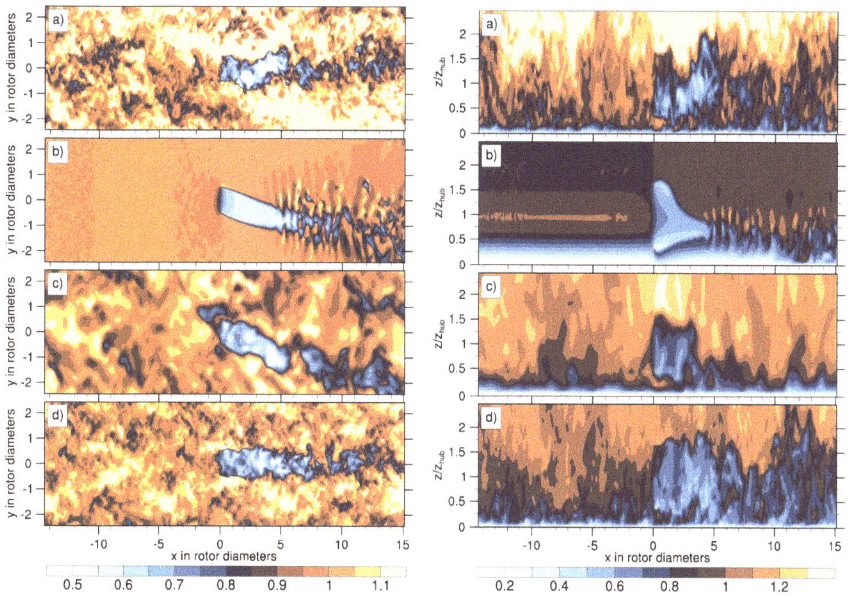

Fig. 3 Snapshots of the horizontal normalized velocity u for different inflow boundary conditions: (a) cyclic, b) non-cyclic, c) non-cyclic with large domain, d) non-cyclic with turbulent inflow). Left: horizontal cross-section in hub height; right: vertical cross-section through hub location.

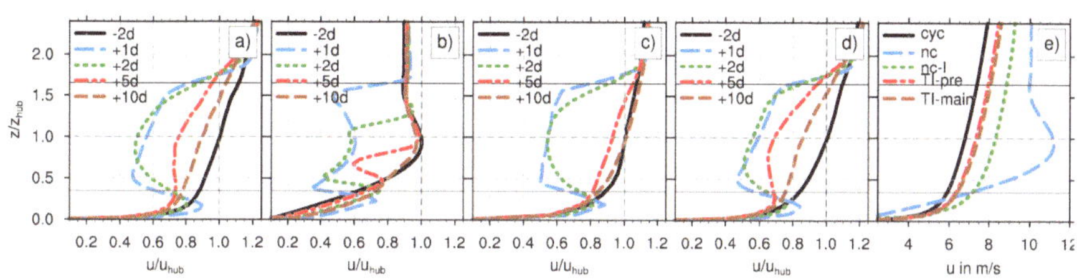

Fig. 4 Normalized vertical profiles of the time-averaged horizontal velocity u upstream and at several locations downstream of the turbine for different inflow boundary conditions: (a) cyclic, b) non-cyclic, c) non-cyclic with large domain, d) non-cyclic with turbulent inflow). e): inflow profiles for the different boundary conditions and the precursor run.

fluctuations of all prognostic variables are extracted from a recycling plane at some distance downstream and added to the mean inflow profiles. This is done at each time step. To preserve the spatial and temporal correlations of the turbulence generated in the precursor run, the distance between inflow and recycling plane should be equal to the domain size of the precursor run. To prevent the boundary layer from growing with increasing distance from the inflow, the recycling is limited to the height of the boundary layer at the inflow. As Fig. 3d) shows, the atmospheric turbulence is well developed already at the inflow. The profile 2 D upstream of the turbine is still similar to that of the precursor run (see Fig. 4e)). Compared to the cyclic simulation, the inflow profile of the simulation with turbulent inflow shows a higher velocity. In the latter case, the flow is undisturbed, whereas in the cyclic simulation the wake deficit of the turbine is being distributed over the whole model domain. This leads to a gradually decreasing inflow velocity in the cyclic simulation. The lower velocity level of the turbulent inflow simulation compared to the laminar simulation with large domain can be explained by the fact that in the laminar case the flow has still not reached a stationary state. At an early simulation time the cyclic run (and precursor run) shows a similar high velocity level. The velocity profiles in Fig. 4a) to d) illustrate once again the inappropriate results of the non-cyclic simulation with small domain.

4 Conclusions

Two turbine models implemented in the LES model PALM have been compared. The simple ADM cannot reproduce the detailed wake structure generated by the sophisticated ALM in the near wake. In the far wake, however, both models produce similar results. This leads to the conclusion that for wind farm simulations with rather large distances between the turbines the ADM is an adequate and cost-saving alternative to the ALM. For wind farms with small turbine spacing the ALM should be used. As this is computationally not feasible for large wind farms, a new method producing similar results as the ALM but saving a lot of computation time is currently being implemented.

Furthermore, the importance of choosing appropriate inflow boundary conditions has been demonstrated. Periodic boundary conditions, as they are commonly used in LES, lead to a modified and no longer undisturbed inflow profile. Non-cyclic boundary conditions with the presented turbulence recycling method guarantee an undisturbed inflow with fully developed atmospheric turbulence and do not require very large model domains.

Acknowledgements. This study has been funded by the German Federal Ministry for the Environment, Nature Conservation and Nuclear Safety (FKZ: 0325220). Computing resources have been partly provided by the North German Supercomputing Alliance (HLRN).

References

1. Calaf, M., Meneveau, C., Meyers, J.: Large-eddy simulation study of fully developed wind-turbine array boundary layers. Phys. Fluids 22, 015110 (2010)
2. Crespo, A., Hernández, J., Frandsen, S.: Survey of Modelling Methods for Wind Turbine Wakes and Wind Farms. Wind Energy 2, 1–24 (1999)
3. Jiménez, A., Crespo, A., Migoya, E.: Application of a LES technique to characterize the wake deflection of a wind turbine in yaw. Wind Energy 13, 559–572 (2010)
4. Kataoka, H., Mizuno, M.: Numerical flow computation around aeroelastic 3D square cylinder using inflow turbulence. Wind and Structures 5, 379–392 (2002)

5. Porté-Agel, F., Wu, Y.T., Lu, H., Conzemius, J.: Large-eddy simulation of atmospheric boundary layer flow through wind turbines and wind farms. J. Wind Eng. Ind. Aerodyn. 99, 154–168 (2011)
6. Prospathopoulos, J.M., Politis, E.S., Rados, K.G., Chaviaropoulos, P.K.: Evaluation of the effects of turbulence model enhancements on wind turbine wake predictions. Wind Energy 14, 285–300 (2011)
7. Raasch, S., Schröter, M.: A large-eddy simulation model performing on massively parallel computers. Meteor. Z 10, 363–372 (2001)
8. Troldborg, N.: Actuator Line Modeling of Wind Turbine Wakes. Dissertation. Fluid Mechanics, Department of Mechanical Engineering, Technical University of Denmark (2008)

Part V

Atmospheric Boundary Layer

Solving 2D Unsteady Turbulent Boundary Layer Flows with a Quasi-Simultaneous Interaction Method

H.A. Bijleveld and A.E.P. Veldman

Abstract. Simulations of a 2D unsteady turbulent boundary layer flow model with a quasi-simultaneous interaction method converge for attached and separated flow. The interaction method ensures that no singularity arises in the numerical system of equations. The singularity will occur if the interaction law coefficient is positive and higher in value than a critical value which depends on the applied flow model. An analysis of the eigenvalues of the system of equations proves this. A sign change in one of the eigenvalues causes the singularity and a proper value of the interaction law coefficient prevents this.

The value of the interaction law coefficient is chosen according to thin-airfoil theory. Simulations of flows over dented plates and airfoils show that this interaction law coefficient suffices to obtain converged results.

1 Introduction

In the RotorFlow project of ECN, the flow over a wind turbine blade is being simulated by splitting the flow domain into a viscous and inviscid region, see fig. 1. By adding viscous effects, a more accurate result can be obtained than with the widely used BEM-methods which are based on a potential flow model and empirical data which results in a fast method in terms of computational time (see e.g. Hansen [4]). However, our method is still fast compared to RaNS computations.

To ensure a smooth connection between the two flow regions a viscous-inviscid interaction method is needed. Here, the quasi-simultaneous interaction method,

---

H.A. Bijleveld
Energy research Centre of the Netherlands, Westerduinweg 3, 1755 ZG Petten, The Netherlands
e-mail: henny.bijleveld.work@gmail.com

A.E.P. Veldman
Institute for Mathematics and Computer Science, University of Groningen, P.O. Box 407, 9700 AK Groningen, The Netherlands
e-mail: a.e.p.veldman@rug.nl

M. Hölling et al. (eds.), *Wind Energy – Impact of Turbulence*, Research Topics in Wind Energy 2,
DOI: 10.1007/978-3-642-54696-9_27, © Springer-Verlag Berlin Heidelberg 2014

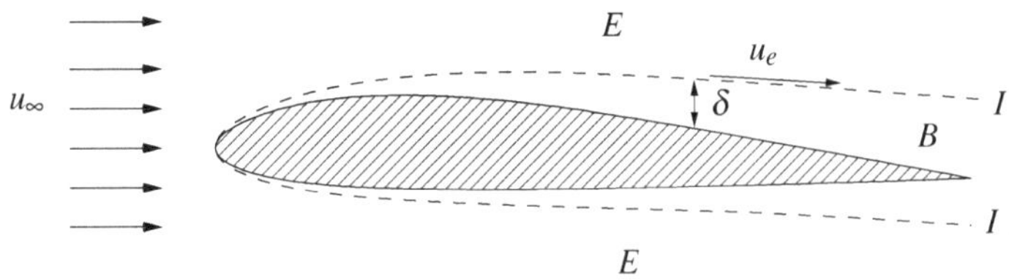

Fig. 1 The flow domains: inviscid external flow (E); viscous boundary layer (B), viscous-inviscid interaction (dashed line, I)

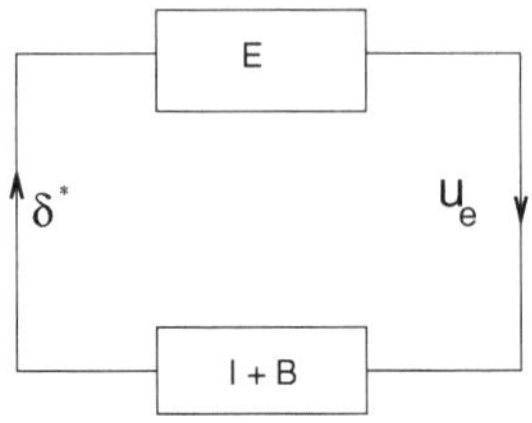

Fig. 2 Quasi-simultaneous interaction scheme with exchange of δ^* and u_e; viscous boundary layer (B), inviscid external flow (E) and interaction law (I)

developed by Veldman [7, 8], is applied. The viscous region is modeled with integral boundary layer equations (B). The set of equations B is composed of von Kármán's momentum integral equation and Head's entrainment equation:

$$\frac{1}{u_e^2}\frac{\partial u_e\delta^*}{\partial t} + \frac{\partial \theta}{\partial x} + \frac{\theta}{u_e}(2+H)\frac{\mathrm{d}u_e}{\mathrm{d}x} = \frac{1}{2}C_f \tag{1}$$

$$\frac{\partial}{\partial t}\left(\delta^*\left(\frac{H_1}{H}+1\right)\right) + \frac{1}{u_e}\frac{\partial}{\partial x}\left(\frac{u_e H_1 \delta^*}{H}\right) = C_E \tag{2}$$

where u_e is the velocity, δ^* the displacement thickness, θ the momentum thickness, $H = \delta^*/\theta$ the shape factor, C_f the skin-friction coefficient, H_1 the turbulent modeled according to Houwink and Veldman [6] and C_E the entrainment coefficient with Head's model [5].

These two equations are solved together with the interaction-law equations (I) after which the external flow equations (E) are solved, see also figure 2. The problem to solve can be written as equation 3.

$$P_{quasi-simultaneous}\left\{\begin{array}{l}\left\{\begin{array}{ll} u_{e_B}^n - I(\delta^n) & = u_{e_E}^{n-1} - I(\delta^{n-1}) \\ u_{e_B}^n - B(\delta^n) & = 0 \end{array}\right. \\ \quad u_{e_E}^n - E(\delta^n) = 0 \end{array}\right. \tag{3}$$

with n the iteration number, u_e the boundary layer edge velocity and δ^* the boundary layer displacement thickness. If the solution is converged, the interaction law drops out of the equation. This means that the choice of interaction law does not influence the converged result, but it does control the convergence of the viscous-inviscid iterations.

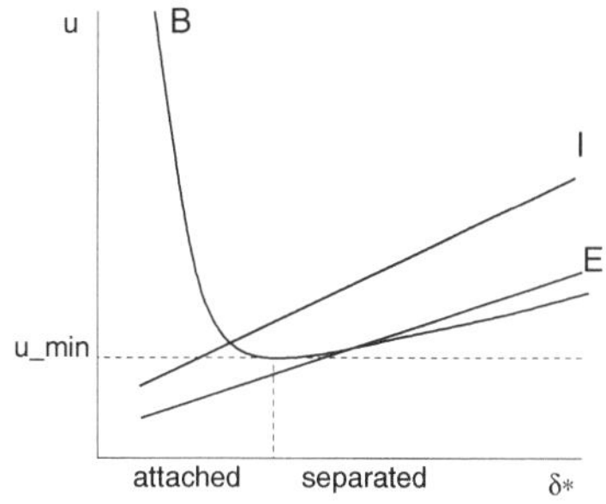

Fig. 3 Relation between viscous (B), inviscid (E) and interaction-law (I) equations as a function of the velocity (u) and boundary layer displacement thickness (δ)

2 Need for Interaction

Interaction methods are applied to ensure a smooth connection between the two flow domains. The interaction can be established in many ways. The most straightforward and easiest is the direct method. Here the velocity is treated as a known from the inviscid calculation and used to calculate the displacement thickness δ^* from the viscous model B (take $I = 0$ in figure 2). The drawback of this method is that at separation a singularity occurs. This corresponds to the mininimum in figure 3. In 2D this is the well-known Goldstein singularity. To avoid this, a (quasi-)simultaneous solution of the external flow is needed. Solving the external and boundary layer flow simultaneously gives a complex method in terms of computer code and flexibility of the method. In a quasi-simultaneous method the boundary layer flow is solved with an *approximation* of the external flow after which the external flow is solved (see also figure 2).

The approximation of the external flow is called the *interaction law* (I) and is an algebraic relation between the velocity and displacement thickness:

$$I : u_e - c\delta^* = RHS(E), \tag{4}$$

with c the interaction law coefficient and $RHS(E)$ information from the external flow. A proper choice of the value of c in equation 4 is important in order to obtain a converging method. The value of c has to be positive enough to ensure an intersection between B and I in figure 3. A minimum value of c is c_{crit} which is an indication of the slope of B. Furthermore, as I is an appproximation of E, the slope of both lines in figure 3 should not deviate too much. From this we can pose two criteria for c:

1. the slope of I is larger than the slope of B: $c > c_{crit}$, with c_{crit} based on equations 1 and 2;
2. the slope of I is close enough to the slope of E (figure 3) for the viscous-inviscid iterations to converge.

The applied method uses $c = \frac{4u_{e,0}}{\pi \Delta x}$ where $u_{e,0}$ is the undisturbed flow and Δx the mesh width. This value is based on a thin-airfoil approximation of the external flow and was also successfully used by Coenen [3] and Bijleveld [2]. From the chosen value of c it follows that for very large mesh sizes the value of c can become too small to ensure a converged result, but for smaller mesh sizes the method works fine.

3 Equations B and I

In the quasi-simultaneous interaction scheme the equations modeling the boundary layer B (eqns. 1 2) are solved together with the interaction law equations I (see also figure 2). The interaction law is used to substitute u_e in B by the interaction law, equation 4. The combined system of equations can be written as:

$$M\frac{\partial \phi}{\partial t} + N\frac{\partial \phi}{\partial x} = R, \quad \phi = \{\delta^*, H\} \tag{5}$$

with

$$M = \begin{bmatrix} \frac{1}{u_e}\left(1+\frac{c\delta^*}{u_e}\right) & 0 \\ \frac{1}{u_e}\left(\frac{H-1}{H}+1\right) & \frac{\delta^*}{u_e H}\left(\frac{\mathrm{d}H_1}{\mathrm{d}H}-\frac{H_1}{H}\right) \end{bmatrix} \tag{6}$$

$$N = \begin{bmatrix} \frac{1}{H}+\frac{c\delta^*}{u_e}\left(1+\frac{2}{H}\right) & -\frac{\delta^*}{H^2} \\ \frac{H_1}{H}\left(1+\frac{c\delta^*}{u_e}\right) & \frac{\delta^*}{H}\left(\frac{\mathrm{d}H_1}{\mathrm{d}H}-\frac{H_1}{H}\right) \end{bmatrix} \tag{7}$$

$$R = \begin{Bmatrix} \frac{C_f}{2} - \frac{\delta^*}{u_e^2}\frac{\partial E}{\partial t} - \frac{\delta^*}{u_e}\left(1+\frac{2}{H}\right)\frac{\partial E}{\partial x} \\ C_E - \frac{H_1\delta^*}{u_e H}\frac{\partial E}{\partial x} \end{Bmatrix} \tag{8}$$

The before mentioned singularity at separation can be related to a sign change of one of the eigenvalues of the steady system of equations, i.e. N. This can happen at separation where experiments show that $\frac{\mathrm{d}H_1}{\mathrm{d}H} = 0$. If the interaction law coefficient is equal to zero ($c = 0$), then N becomes singular.

The criterion $c > c_{crit}$ takes care that no sign change occurs. The value of c_{crit} is the lowest value of c for which no sign change in the system of equation 5 occurs: $\det(NM^{-1}) > 0$.

Because it is impossible for M to become singular, we get c_{crit} from $\det(N) = 0$. Note that for separated flow $c_{crit} > 0$, hence the direct method with $c = 0$ fails (Goldstein singularity).

4 Numerical Results

Simulations of boundary layer flow over an airfoil and a dented plate have been performed in order to test the method. For the viscous calculations, the discretization scheme is a first order upwind scheme. The direction of discretization depends on the value of the eigenvalues of the system $(B+I)$, equation 5. The external flow is modeled as a potential flow and solved with a panel method.

4.1 Airfoil

Simulations of attached boundary layer flow over a NACA0012 airfoil have been performed. Table 1 shows the lift coefficient for a small range of angles of attack. The results correspond nicely with the experimental data.

Table 1 c_l-α data for flow over a NACA0012 airfoil, $Re = 9.0 \cdot 10^6$, 121 mesh points. The lift coefficients between brackets are experimental data [1].

$\alpha(°)$	$c_l(-)$
0.0	0.0
2.0	0.23 (0.21)
4.0	0.45 (0.44)
6.0	0.68 (0.67)
8.0	0.90 (0.89)
10.0	1.11 (1.09)

4.2 Dented Plate

Simulations of flows over a dented plate with 5% dent depth have been performed. In the dent, separation occurs. Figure 4 shows the eigenvalues of $M_{-1}N$ from equation 5 along the plate (left) and interaction law coeffcients (right).

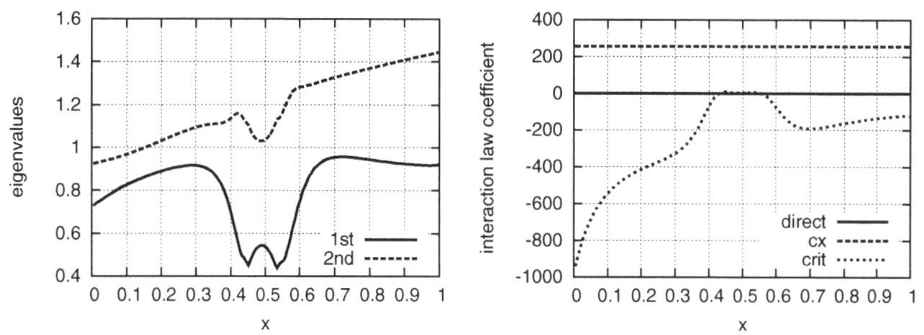

(a) Eigenvalues (b) Critical value of c

Fig. 4 Eigenvalues (left) and critical value of c (right) for a 2D unsteady turbulent boundary layer flow over a dented plate with 5% dent depth. $Re = 11.5 \cdot 10^6$; $U_\infty = 1$; $\Delta x = 1/200$.

In figure 4a it can be seen that the eigenvalues remain positive on all mesh points, even in the separated region. Note that also the eigenvalues are relatively low in this region.

In the right plot of figure 4, the interaction coefficient c is plotted for the current case (`cx`), a direct mode (`direct`) and the critical value (`crit`). The region with positive c_{crit} in figure 4b corresponds to separated flow. The figure shows that everywhere: $cx > c_{crit}$, thus fulfilling the first criterion for c.

5 Conclusions

The quasi-simultaneous interaction method is a robust method for application in simulations of turbulent boundary layer flow where the flow can be both attached and separated. The interaction law is an approximation of the external flow and solved together with the turbulent boundary layer equations.

A proper value of the interaction law coefficient in the interaction law equations prevents the system from becoming singular at separation. The value should be positive and larger than a certain value that follows from the chosen boundary layer model ($c > c_{crit}$).

Analysis of the system of equations and simulations with unsteady turbulent flow over dented plates and airfoils show that there is no sign change in the eigenvalues. Simulations also show that the interaction coefficient meets the requirements. From this we can draw the conclusion that the method works.

References

1. Abbott, I.H., von Doenhoff, A.E.: Theory of Wing Sections. Dover, New York (1959)
2. Bijleveld, H.A., Veldman, A.E.P.: Quasi-simultaneous interaction method for solving boundary layer flows in primary and characteristic variables. In: 7th PhD Seminar on Wind Energy, EWEA, Delft, NL, October 27-28 (2011)
3. Coenen, E.G.M.: Viscous-Inviscid Interaction with the Quasi-Simultaneous Method for 2D and 3D Aerodynamic flow. PhD thesis, Rijksuniversiteit Groningen, the Netherlands (2001)
4. Hansen, M.O.L.: Aerodynamics of Wind Turbines, 2nd edn. Earthscan, London (2008)
5. Head, M.R.: Entrainment in the turbulent boundary layer. Aeronautical Research Council R&M 3152 (1960)
6. Houwink, R., Veldman, A.E.P.: Steady and Unsteady Flow Computations for Transonic Airfoils. AIAA Journal 84, 1618 (1984)
7. Veldman, A.E.P.: New, quasi-simultaneous method to calculate interaction boundary layers. AIAA Journal 19, 79–85 (1981)
8. Veldman, A.E.P.: A simple interaction law for viscous-inviscid interaction. Journal of Engineering Mathematics 65, 367–383 (2009)

Time - Resolved CFD Simulation of a Turbulent Atmospheric Boundary Layer Interacting with a Wind Turbine

K. Meister, Th. Lutz, and E. Krämer

Abstract. The present article describes results of an unsteady CFD simulation of the generic 5MW NREL turbine under unsteady atmospheric inflow conditions. Thereby, the blade loads are evaluated and the wake development is investigated.

1 Introduction

Wind turbines in operation are exposed to the turbulent atmospheric boundary layer. The boundary layer is characterized by large scale eddies, shear flow and velocity variation over time and height. These characteristics strongly affect the blade loads as the relative angle of attack changes permanently. The change in angle of attack causes unsteady nonlinear blade load response in the linear region of originally steady aerodynamic polars which should be considered for reduced frequencies higher than 0.04 [1]. Further the large scale turbulence interacts with the turbine wake and the wake begins to meander [2].

Commonly used for wind turbine simulations in an atmospheric boundary layer is the Actuator Line Method in combination with a LES [3] [4] which uses two dimensional airfoil data in combination with 3d models to determine the three dimensional turbine flow. However, due to the fact of modeling when using the Actuator Line Method only CFD methods calculating directly the unsteady pressure distribution can consider unsteady three dimensional and nonlinear effects physically correct. For this purpose the Institute of Aerodynamics and Gas Dynamics (IAG) performs CFD simulations of fully meshed wind turbines in an unsteady atmospheric boundary layer. The results should be used to better predict turbine loads and improve the understanding of the near wake flow and the interaction of turbines with atmospheric

K. Meister · Th. Lutz · E. Krämer
Institute of Aerodynamics & Gas Dynamics (IAG), University of Stuttgart,
Pfaffenwaldring 21 70569 Stuttgart
e-mail: {meister,lutz,kraemer}@iag.uni-stuttgart.de

M. Hölling et al. (eds.), *Wind Energy – Impact of Turbulence*,
Research Topics in Wind Energy 2,
DOI: 10.1007/978-3-642-54696-9_28,

turbulence. In doing so this article describes the numerical model and simulation results of the generic NREL 5MW turbine that has been used as reference turbine in the European UpWind and is improved in the ongoing OFFWINDTECH project [5].

2 Description of the Numerical Model

The simulations are hybrid RANS / LES simulations (DES) and are performed by using the CFD solver FLOWer, which is provided by the German Aerospace Center (DLR) and was successfully used for wind turbine simulations during the MexNext project [6]. DES means, that the blade boundary layer is calculated in RANS mode with statistical turbulence modeling. The atmospheric boundary layer is simulated in LES mode with temporal and spatial resolution of the relevant large scale turbulent structures, except a thin wall layer where the used DES 97 method switches to RANS. In comparison to prior simulations [7] the time step is reduced to 0.04 s ($\approx 2.9^{\circ}$), the distance from the inflow to the turbine location is set to 250 m and the wake is refined.

Geometry of the Reference Turbine

The 5MW NREL turbine is a three bladed turbine with a rotor diameter of 126 m and a hub height of 90 m. The rotor plane is tilted by 5° while the blades show a precone angle of -2.5°. The maximum tip speed is 80 m/s and correlates with the maximum rotor speed of 12.1 rpm. While the rotor geometry was described detailed in the European UpWind project hub and nacelle need to be approximated. The hub diameter was determined to 4.4 m while the nacelle was approximated with a cuboid of $18\,\text{m} \times 6\,\text{m} \times 6\,\text{m}$.

Used Meshes

The complete computational setup consists of several different meshes which are listed in Tab. 1. As the blades are the most important aerodynamic structures of a wind turbine the use of high quality meshes is mandatory. Therefore, a dedicated blade mesh generation script developed at the IAG is used [8].

All other meshes are created manually with a fully resolved boundary layer. The background mesh has a grid resolution of 2.5 m in all three directions of space except at ground level where the cell spacing increases from very low cell heights up to 2.5 m to enable a fully resolved boundary layer with $y^{+} \approx 1$. The designed grid provides accuracy for the RANS region at ground level, and good efficiency for the LES region of the background mesh. It covers a flow region of $760\,\text{m} \times 660\,\text{m} \times 200\,\text{m}$.

Unsteady Atmospheric Inflow Data

To consider unsteady atmospheric inflow data a Dirichlet boundary condition being developed during the German OWEA project was used. This boundary condition feeds inflow data of already existing simulation result of a pure turbulent atmosphere

Table 1 Overview of the meshes used for the simulation of the reference turbine

Mesh	Cells (mio)	No.	Description
Blade	2	3	Rotating mesh with fully resolved boundary layer $y^+ \approx 1$
Hub	2.5	1	Rotating mesh with resolved boundary layer.
Nacelle	1.1	1	Steady mesh with resolved boundary layer.
Tower	0.6	1	Steady mesh with resolved boundary layer.
Tip vortex	4.2	1	Steady mesh covering the blade tip region behind the turbine for better vortex resolution
Background	9.6	1	Steady mesh covering the environment around the turbine.
Auxiliary	0.8	-	Several smaller rotating and steady meshes building connections to other meshes

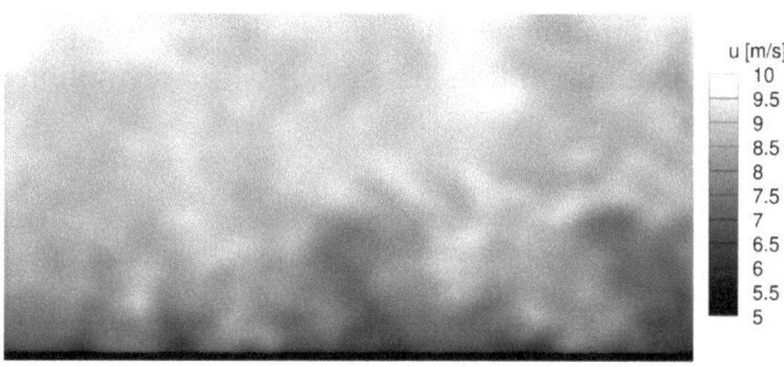

Fig. 1 Extract of LES data provided by DEWI; The data were extracted in a yz-plane at a specific time step; Several time steps are used as inflow data for the simulations; The data provides velocities up to 10 m/s and has yaw included

into the simulation far upstream of the turbine. For the present simulation highly resolved LES data from the German Wind Energy Institute (DEWI) were used as turbulent atmospheric inflow data. These data were extracted from a prior LES simulation of a maritime atmospheric boundary layer using the CFD code PALM and contain the wind components u, v and w with a resolution of $\Delta x \times \Delta y \times \Delta z = 5\,\mathrm{m} \times 5\,\mathrm{m} \times 5\,\mathrm{m}$ [9] (Fig. 1).

3 Simulation Results

To analyze the influence of turbulent atmospheric boundary layer two simulations with different inflow data were performed using the described method. While simulation I used a uniform inflow, simulation II was performed with the prior described unsteady atmospheric inflow data. The constant inflow velocity of simulation I was chosen to fit the maximum velocity at hub height of the unsteady inflow data.

Axial Blade Loads

The evaluation of the rotor axial blade load force F_{ax} was performed by integration of the aligned blade pressure distribution for each rotor blade. Figure 2 (a) shows the rotor axial blade load F_{ax} for blade 1 in uniform inflow (black solid line) compared to blades 1-3 in unsteady inflow. Significant is the overall load reduction in unsteady inflow caused by the velocity deficit in the atmospheric boundary layer. To analyze the differences between uniform and turbulent inflow the rotor axial blade force is normalized with its mean value (figure 2 (b)). The resulting load distribution for uniform inflow shows almost circular shape. Only during tower passage (120° - 200°) the loads show stronger variation. The loads are slightly higher for the blade approaching the towerthen for the blade leaving the tower region. Moreover, $\bar{F}_{ax}$ shows little variation up to 2% over one revolution for the uniform inflow while this variation could not be observed for smaller pitch angles. The simulation with turbulent inflow results in higher loads in the upper region of the rotor disc. This correlates with the higher inflow velocities of the boundary layer in the upper region. The loads differ between $+8\%$ and -13% and show no circular shape due to the permanent changing inflow. As the load evaluation was performed over one full rotation (0° - 360°) and the inflow data covers $4.0\bar{3}$ revolutions a jump can be detected at 0°.

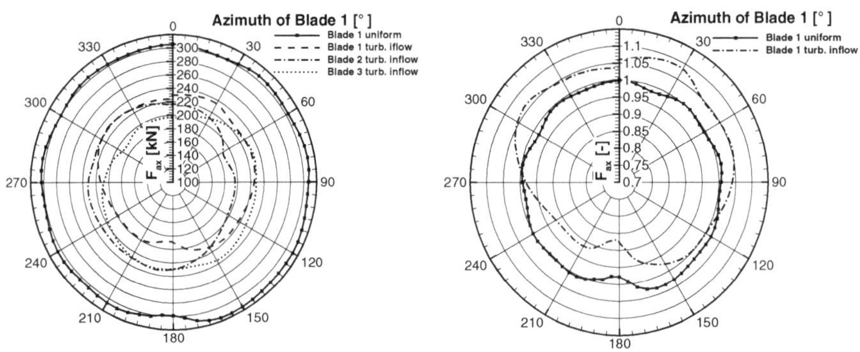

(a) Rotor axial blade force F_{ax} (b) Normalized rotor axial blade force $\bar{F}_{ax}$

Fig. 2 Comparison of rotor axial blade load F_{ax} under uniform inflow (10 m/s) and unsteady inflow data from DEWI (10 m/s peak at hub height). Extract of one revolution is shown.

Near Wake Movement

The movement of the wake is influenced by shear, yaw and turbulence of the upstream flow. To determine the wake movement the wake center is analyzed. This is done by calculating the angular momentum center of the wake

$$Y_{Center} = \frac{\int y \cdot (u_\infty - u_{ref}) \rho \, \mathrm{dxdydz}}{\int (u_\infty - u_{ref}) \rho \, \mathrm{dxdydz}} \tag{1}$$

with the assumption of $\rho = \text{const}$. To reduce the influence of the boundary layer in the wake center calculation velocities higher than 8.6 m/s are ignored in equation 1 as well as velocities which occur at z positions lower than 10 m. Using this evaluation method the wake center can be determined (Fig. 3). Analyzing and tracking

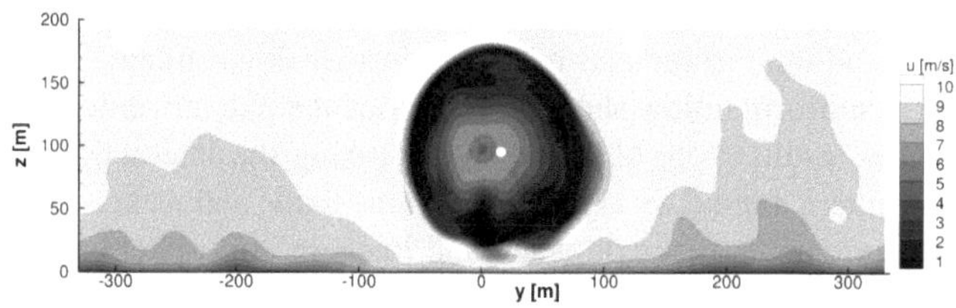

Fig. 3 Velocity u in yz-plane at 60 m downstream the turbine with wake center (white dot)

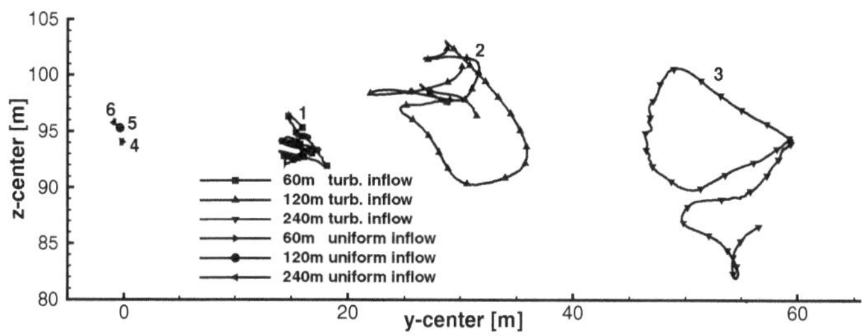

Fig. 4 Wake movement over time at three downstream positions of a yz-plane for the turbulent inflow (line 1-3) and the uniform inflow (line 4-6)

the wake center position over time the wake movement can be determined (Fig. 4). The wake centers were tracked for several x-constant planes at 60 m, 120 m and 240 m downstream position. The tracks show the different wake movement for turbulent inflow (1-3) and uniform inflow (4-6). The rotor center for the uniform inflow (evaluated over two revolutions) is more or less constant unlike the wake center for the turbulent inflow. Due to the yaw of the inflow data the wake is shifted. Moreover it varies in horizontally and vertically. This movement increases when moving downstream. Even though the wake with turbulent inflow was evaluated over eight revolutions the wake center movement is not periodic.

4 Conclusion

This article presents unsteady CFD results of the generic 5MW NREL turbine in time-resolved turbulent atmospheric inflow. The performed DES simulation used meshes with fully resolved boundary layer for the blades and the turbine. The turbulent inflow data were extracted from a prior LES simulation and propagated time resolved to the turbine. Analyzing the blade loads the influence of the boundary layer shear and the turbulence could be shown as well as the non periodic load response due to changing flow conditions. While simulations of the turbine in uniform inflow showed load fluctuations of 2% the simulation in turbulent atmospheric inflow revealed up to 13% fluctuation. Moreover, the wake movement was analyzed by calculating the wake center. It could be investigated that the wake movement begins right behind the turbine and increases further downstream for the turbulent inflow while the uniform inflow shows no significant wake movement.

The presented results were obtained by analyzing the interaction of a specific turbulent atmospheric boundary layer with turbine loads and wake.

Current measurements focus on wake and loads in atmospheric boundary layer but without time accurate correlation to the transient inflow. For future simulations and validation more measurements considering this topic are desired.

Acknowledgements. The presented studies were part of the RAVE project "OWEA". The authors wish to acknowledge to "German Federal Ministry for the Environment, Nature Conservation and Nuclear Safety for the funding of this research.

References

1. Piziali, R.A.: 2-D and 3-D Oscillating Wing Aerodynamics for a Range of Angels of Attack Including Stall. Technical report, Ames Research Center - NASA (September 1994)
2. Larsen, G.C., Madsen, H., Bingöl, F.: Dynamic wake meandering modeling. Risø-R-Report (June 2007)
3. Jimenez, A., Crespo, A., Migoya, E., Garcia, J.: Advances in large-eddy simulation of a wind turbine wake. The Science of Making Torque from Wind (2007)
4. Calaf, M., Meneveau, C., Meyers, J.: Large eddy simulation study of fully developed wind-turbine array boundary layers. Physics of Fluids 22 (January 2010)
5. Jonkman, J., Butterfield, S., Musial, W., Scott, G.: Definition of a 5-MW Reference Wind Turbine for Offshore System Development. Technical Report NREL/TP-500-38060 (February 2009)
6. Schepers, J.G., Boorsma, K., Bon, A.: Results from Mexnext: Analysis of detailed aerodynamic measurements on a 4.5m diameter rotor placed in the large German Dutch Wind Tunnel DNW. In: Proceedings of the European Wind Energy Conference, EWEC (2011)
7. Meister, K., Lutz, T., Krämer, E.: Consideration of unsteady inflow conditions in wind turbine CFD simulations. In: Proceedings of the German Wind Energy Conference, DEWEK (2010)
8. Meister, K., Lutz, T., Krämer, E.: Development of a process chain for a detailed wake simulation of horizontal axis wind turbines. In: EUROMECH[508] - Wind Turbine Wakes, Madrid, October 20-22 (2009)
9. Cañadillas, B.: A study of the marine boundary layer by LES-modeling and experimental observations with a focus on offshore wind energy applications. PhD Thesis, Hannover (2009)

Mesoscale Modeling of Low-Level Jets over the North Sea

Christopher Nunalee and Sukanta Basu

Abstract. Contemporary onshore and offshore wind resource assessment frameworks incorporate diverse multi-scale weather prediction models (commonly known as mesoscale models) to dynamically downscale global-scale atmospheric fields to regional-scale (i.e., spatial and temporal resolutions of a few kilometers and a few minutes, respectively). These high resolution mesoscale models aim at depicting the expected wind behavior (e.g., wind shear, wind turning, topographically induced flow accelerations) at a particular location or region. Over the years, numerous model sensitivity and intercomparison studies have investigated the strengths and weaknesses of the models' parameterizations (including, but not limited to, planetary boundary layer turbulence) in capturing realistic flows over land. In contrast, only a handful of modeling studies have focused on coastal and offshore flows (e.g., coastal fronts, internal boundary layers, land breeze - sea breeze circulations, low-level jets); thus, our understanding and predictive capability of these flows remain less than desirable. This impairment, in combination with the recent world-wide surge in offshore wind energy development, provides the rationale for this study.

We are currently evaluating the performance of the Weather Research and Forecasting (WRF) model, a new-generation mesoscale model, in simulating some of the aforementioned coastal and offshore flow phenomena. In this paper, we focus on low-level jets and compare the WRF model-simulated results against the observational data from the FINO1 meteorological mast in the North Sea. We also discuss the sensitivities of the WRF model-generated offshore wind fields with respect to several planetary boundary layer turbulence schemes.

1 Introduction

In comparison to inland areas, coastal and offshore regions are more attractive from a wind energy development perspective, for a number of physical and practical reasons: favorable wind resource due to reduced surface friction over water; close

Christopher Nunalee · Sukanta Basu
Department of Marine, Earth and Atmospheric Sciences, North Carolina State University, Raleigh, NC, 27695, USA
e-mail: {cgnunale, sukanta_basu}@ncsu.edu

M. Hölling et al. (eds.), *Wind Energy – Impact of Turbulence*,
Research Topics in Wind Energy 2,
DOI: 10.1007/978-3-642-54696-9_29, © Springer-Verlag Berlin Heidelberg 2014

proximity to large population centers; relative ease of transportation and installation of large turbines; etc. [5, 6, 9]. Thus, it is not surprising that many countries around the world have been rapidly increasing their offshore wind energy portfolio in the recent years. In 2011, Europe installed over 800 MW of offshore wind energy bringing its total capacity to 3,812.6 MW; nearly a 400% increase from 2007 [14].

For offshore wind energy projects to be profitable and sustainable, highly accurate wind resource maps at turbine hub-heights (on the order of 100 m) are essential. However, the challenge to produce these maps is complicated by the fact that our current coastal and offshore flow modeling capabilities are fundamentally limited. The situation is further aggravated by the lack of tall meteorological towers in offshore waters. As a result, most of the mesoscale model-generated offshore wind resource maps produced to date have only been validated against near-surface observations (measured by ocean buoys or estimated by satellite remote sensing) instead of hub-height observations. In the coastal and offshore regions, quite often (e.g., in the presence of internal boundary layers, low-level jets), the near-surface winds are (partially) decoupled from the upper layer winds. Under these scenarios, the validation of mesoscale model results using surface observations is futile – unquestionably, one needs reliable hub-height wind measurements for appropriate validations. In this paper, we document such a model validation study utilizing a new-generation mesoscale model, called the Weather Research and Forecasting (WRF) model [10], and observational data from the state-of-the-art FINO1 mast in North Sea.

The focus of this paper is on the modeling aspects of offshore low-level jets (LLJs), which are relatively narrow ($\sim$100 km) channels of wind maxima typically bounded vertically between $\sim$100 m and $\sim$1000 m above ground level (AGL). They can be caused by various physical mechanisms: inertial oscillations, baroclinicity due to land-sea temperature contrasts, etc. [2, 11]. Over land, nocturnal LLJs are sometimes responsible for high wind resources (see [12] and the references therein); at the same time, they can also have detrimental effects (related to fatigue loading) on wind turbines [4]. The impact of offshore LLJs on wind resources and/or turbines is currently not known in the literature but beyond the scope of this study.

2 Data and Methodology

In this study, three consecutive offshore LLJ events (May 4-7, 2008) are identified based on the observed wind data from the FINO1 offshore meteorological mast (Figure 1 – right panel). On this mast, wind speed and direction values are measured at multiple levels between 33 m and 100 m AGL with a temporal resolution of 10 minutes [8]. These data are used to assess the accuracy of various WRF model (version 3.3.1) runs while simulating the identified LLJ events.

The WRF model's preprocessing system is used to construct a nested modeling domain centered on the location of the FINO1 mast (Figure 1 – left panel). This domain configuration consists of a large parent domain with 2 nested domains. The largest domain employs a 18 km $\times$ 18 km horizontal grid resolution; the intermediate

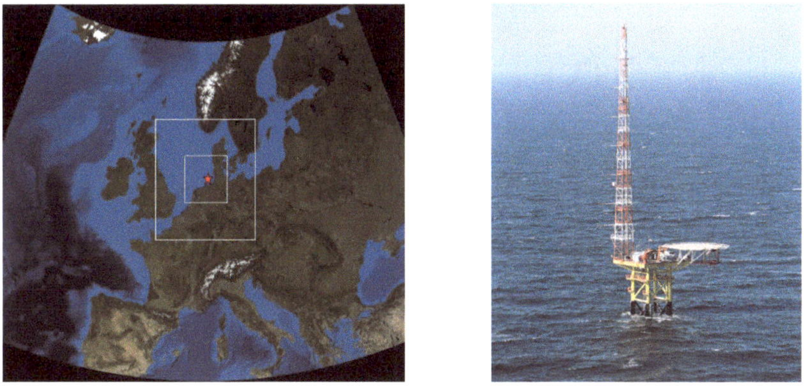

Fig. 1 Mesoscale modeling domains and location of the FINO1 tower (left); photograph of the FINO1 meteorological tower (right). Image courtesy: http://www.dewi.de/

domain uses a horizontal grid resolution of 6 km × 6 km; and the innermost domain uses a horizontal grid resolution of 2 km × 2 km. Throughout all of the domains, the vertical resolution remained constant with 50 grid points between the surface and ∼ 16,000 m AGL; the lowest model level is at ∼8 m AGL; and 18 out of the 50 total vertical grid points are below 1 km AGL.

In order to evaluate the performance of various planetary boundary layer (PBL) schemes, the WRF model is run multiple times. For each model run, a different PBL scheme is invoked while all other model configurations are held constant. The various PBL schemes evaluated here are summarized in Table 1.

Table 1 Summary of the Specific PBL Schemes Evaluated

PBL Scheme	Reference
Mellor-Yamada-Janjič (MYJ)	Janjič [3]
Mellor-Yamada Nakanishi and Niino Level 2.5 (MYNN2)	Nakanishi and Niino [7]
Quasi-Normal Scale Elimination (QNSE)	Sukoriansky et al. [13]
Yonsei University (YSU)	Hong et al. [1]

3 Results

Wind speed observations from the FINO1 mast at multiple heights (Figure 2 - left panel) display three consecutive diurnal cycles. The peaks of the cycles represent three low-level jets. These jets appear to be forming around 18 Z (early evening) and disappearing around 4 Z (early morning). Strong wind speeds (∼ 14 m s^{-1} at 100 m height) occur at the times when the LLJs are the most prominent. All the WRF model runs qualitatively capture these behaviors. However, the diversity among the

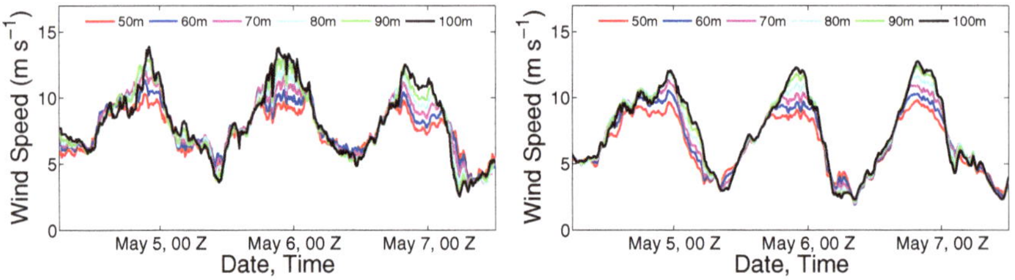

Fig. 2 Wind speed time-series from the FINO1 mast (left panel) and a WRF model run using the QNSE PBL scheme (right panel). Observed and simulated series clearly portray three diurnal cycles. The peaks of the cycles correspond to the LLJ events.

WRF runs utilizing different PBL schemes are significant. Figure 3 illustrates the broad distribution of wind speed intensities within each jet core, with the first and third jets showing the greatest degree of variability. With regards to the third jet, the MYJ and QNSE PBL schemes produce the most intense jets with maximum wind speeds around 15 m s^{-1}; on the other hand, the YSU PBL scheme simulates the weakest one with a peak wind speed of less than 10 m s^{-1} in a barely noticeable LLJ structure. In addition to these intensity differences, overall jet architectures vary as well along with the relative heights of the jet peaks.

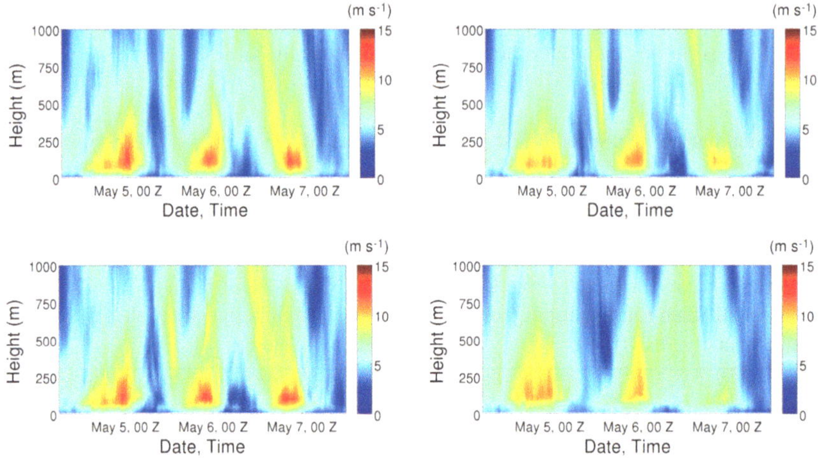

Fig. 3 The WRF model-derived time/height plots of wind speed (m s^{-1}) during May 4-7, 2008 at FINO1. Each panel represents a WRF model run using a specific PBL scheme: MYJ (top left), MYNN2 (top right), QNSE (bottom left), and YSU (bottom right).

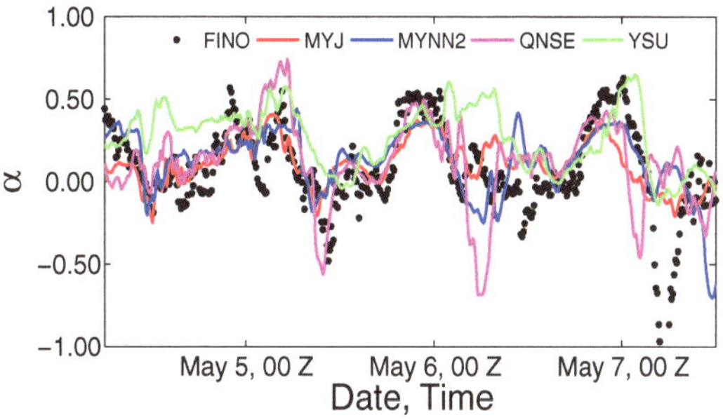

Fig. 4 Observed and the WRF model-derived wind shear exponents corresponding to 50-100 m AGL

Figure 4 displays the temporal evolution of the shear exponent (α). It is commonly defined as:

$$\left(\frac{U_{z_2}}{U_{z_1}}\right) \sim \left(\frac{z_2}{z_1}\right)^{\alpha} \tag{1}$$

where U_{z_1} and U_{z_2} are wind speeds at the lower (z_1) and upper (z_2) sensor (or model vertical grid) levels, respectively. In wind resource assessment projects, $\alpha = \frac{1}{7}$ is commonly used. However, from a meteorological standpoint, α is expected to vary with atmospheric stability and surface roughness (see [12] and the references therein). It might also depends on advection and non-equilibrium conditions, which are common in coastal zones [2]. Figure 4 is inline with these expectations. The observed as well as the modeled results show large shear exponents in the presence of LLJs; albeit the model-generated values are smaller than their observed counterparts.

4 Concluding Remarks

It was found that the WRF model qualitatively captures various traits of the observed offshore low-level jets. However, quantitative evaluations reveal departures from observations that are considered to be significant from a wind energy perspective. Additionally, LLJ intensity, structure and height were observed to be especially sensitive to PBL parameterizations. On the other hand, the timing of each jet's development and termination is accurately simulated by all the PBL schemes. Our results emphasize the need for multiphysics ensembles in future offshore wind resource assessment projects.

Acknowledgements. We are grateful to the BMU (Bundesministerium fuer Umwelt, Federal Ministry for the Environment, Nature Conservation and Nuclear Safety) and the PTJ (Projekttraeger Juelich, project executing organisation) for providing us access to the FINO datasets. The computational resources are generously provided by Renaissance Computing Institute. This material is based upon work supported by the National Science Foundation (NSF) under Grant No. AGS-1122315. Any opinions, findings, and conclusions or recommendations expressed in this material are those of the authors and do not necessarily reflect the views of the NSF.

References

1. Hong, S.-Y.: A new stable boundary-layer mixing scheme and its impact on the simulated East Asian summer monsoon. Q. J. R. Meteorol. Soc. 136, 1481–1496 (2010)
2. Hsu, S.A.: Coastal Meteorology. Academic Press, Inc., San Diego (1988)
3. Janjiĉ, Z.I.: The step-mountain eta coordinate model: further developments of the convection, viscous sublayer, and turbulence closure schemes. Monthly Weather Review 122, 927–945 (1994)
4. Kelley, N.D., Shirazi, M., Jager, D., Wilde, S., Sullivan, P., Patton, E.: Lamar low-level jet project interim report. Technical Report NREL/TP-500-34593, National Renewable Energy Laboratory (2004)
5. Lange, B., Larsen, S.E., Hoejstrup, J., Barthelmie, J.: The influence of thermal effects on the wind speed profile of the coastal marine boundary layer. Boundary Layer Meteorology 112, 587–617 (2004)
6. Lavagnini, A., Semprevia, A.M., Barthelmie, R.J.: Estimating wind energy potential offshore in Mediterranean areas. Wind Energy 6, 23–34 (2003)
7. Nakanishi, M., Niino, H.: An improved Mellor-Yamada level-3 model: its numerical stability and application to a regional prediction of advection fog. Boundary Layer Meteorology 119, 397–407 (2006)
8. Neumann, T., Nolopp, K., Herklotz, K.: First operating experience with the fino1 research platform in the north sea. DEWI-Magazin. 24, 27–34 (2004)
9. Semprevia, A.M., Barthelmie, R.J., Pryor, S.C.: Review of methodologies for offshore wind resource assessment in European seas. Surveys in Geophysics 29, 471–497 (2008)
10. Skamarock, W.C., Klemp, J.B., Dudhia, J., Gill, D.O., Barker, D.M., Duda, M.G., Huang, X.-Y., Wang, W., Powers, J.G.: A description of the advanced research wrf version 3. Technical Report NCAR/TN-457+STR, National Center for Atmospheric Research (2008)
11. Smedman, A., Högström, U., Bergström, H.: Low level jets - a decisive factor for offshore wind energy siting in the Baltic Sea. Wind Engineering 28, 137–147 (1996)
12. Storm, B., Basu, S.: The WRF model forecast-derived low-level wind shear climatology over the United States Great Plains. Energies 3, 258–276 (2010)
13. Sukoriansky, S., Galperin, B., Perov, V.: Application of a new spectral theory of stably stratified turbulence to the atmospheric boundary layer over sea ice. Boundary-Layer Meteorology 117, 231–257 (2005)
14. Wilkes, J., Moccia, J., Arapogianni, A., Dragan, M., Plytas, N., Genachte, A.-B., Guillet, J., Wilczek, P.: The european offshore wind industry key 2011 trends and statistics (2012)

Aerodynamic Boundary Layer Investigation on a Wind Turbine Blade under Real Conditions

Daniela Schwab, Stefan Ingwersen, Alois Peter Schaffarczyk, and Michael Breuer

Abstract. For the design and optimization of wind turbine blades knowledge about the state of the boundary layers at the rotor is important since the energy yield strongly depends on this issue. Blades with an extended laminar boundary layer zone offer a higher energy yield. Owing to the complexity of the experimental set-up, up to now most research on transition was conducted in wind tunnels and field measurements are rare. In the present study the aerodynamic boundary layer at a rotor blade is investigated while the turbine is working under real operating conditions in the atmosphere. Hence important effects such as the rotation of the rotor and the unsteady behavior of the inflow are taken into account. The blade is equipped with a hot film and pressure tubes in order to detect disturbances in the boundary layer and to gain information about the flow field. A preliminary analysis of the time signal and the calculated energy spectra of the hot film measurements leads to two results. First, it is possible to determine the state of the boundary layer (laminar or turbulent). Second, we assume that in case of a medium rotational speed transition following the Tollmien-Schlichting scenario occurs. This assumption has to be verified by comparison with further measurements.

1 Introduction

In order to design wind turbines, knowledge of the flow field including the boundary layer behavior at the rotor blade is important. The energy yield of a wind turbine

Daniela Schwab · Stefan Ingwersen · Alois Peter Schaffarczyk
University of Applied Sciences Kiel, Grenzstraße 3, 24149 Kiel, Germany
e-mail: {daniela.schwab, stefan.ingwersen, alois.schaffarczyk}@fh-kiel.de

Michael Breuer
Helmut-Schmidt-University Hamburg, Holstenhofweg 85, 22043 Hamburg, Germany
e-mail: breuer@hsu-hh.de

M. Hölling et al. (eds.), *Wind Energy – Impact of Turbulence*,
Research Topics in Wind Energy 2,
DOI: 10.1007/978-3-642-54696-9_30,

strongly depends on the boundary layer state. Blades, whose boundary layers have an extended laminar zone, offer a higher energy yield. In-flight measurements at airplane profiles show that transition, following different scenarios, may occur.

Due to the lack of detailed experimental results for the flow field, simplified and empirical methods are typically used for the performance prediction of wind turbine and blade design. Up to now most transition predictions have been realized with the so-called e^N method for transition, which takes place after the Tollmien-Schlichting (TS) scenario. A stability criterion is defined using Mack's correlation [2], which was developed for wind tunnel tests and has neither been validated nor modified for atmospheric flow conditions, which significantly differ from wind tunnel data. Hence important effects such as the rotation of the rotor and the unsteady behavior of the inflow are not taken into account. As the experimental set-up for transition measurements is rather complex, most experimental research studies on this topic were conducted in wind tunnels. Thus field measurements under natural conditions are rare. For this reason in the present study a blade of a wind turbine is equipped with a hot-film and pressure tubes in order to investigate the aerodynamic boundary layer and its state while the turbine is working under ordinary operating conditions.

2 Aerodynamics and Transition Scenarios

This section covers a short overview of possible transition scenarios and the e^N method. The above mentioned *natural transition* after the TS scenario follows different steps which have been described in detail by White [9] at the example of a flat plate. Near the leading edge there is a stable laminar flow. The first instability occurs after a certain length in form of two-dimensional waves. These are the so-called TS waves, which in case of amplification develop to unstable three-dimensional waves and vortices. These vortices later break down to turbulence spots which pass on to a fully turbulent flow. For transition predictions the stability of the TS waves, which denote the beginning of the process, must be considered. This is done by the e^N method [2, 6], which is based on the linear stability theory relying on the Orr-Sommerfeld equation. It evaluates the development of disturbances in the boundary layer. The amplification rate N for disturbances with different frequencies is calculated for each position at the profile. The critical N-factor, for which transition occurs, depends on the inflow turbulence and is commonly estimated using Mack's correlation [2]. As already mentioned, this correlation is limited to wind tunnel flows. Furthermore, the linear stability theory also allows to estimate the frequencies of the most strongly amplified disturbances. Another important transition scenario is the *bypass transition*, which takes place for high inflow turbulence. In this scenario the steps with the unstable waves are bypassed, so that streaks and turbulent spots are directly formed. In case of natural transition after the TS scenario the energy spectra should show a characteristic maximum at the frequency range of the disturbances with the strongest amplification. On the contrary, the energy spectra should possess no maxima for transition following the bypass scenario, so that this difference can be used to interpret the hot film measurements [3, 4, 5, 7].

3 Measurements

In the present study the blade of an E30 wind turbine from Enercon GmbH with a rotor diameter of 30 m is equipped with a hot film and pressure tubes in order to investigate the boundary layer and especially the transition process on the blade. The measurement devices are installed at a rotor diameter of 8 m and are shown in Fig. 1. The measurement set-up is based on the in-flight measurement of Seitz [7]. A hot film is attached at the upper side midspan of the blade, covering a profile depth from 25 to 34%. Since the hot film is placed directly on the surface, wall shear stress fluctuations can be measured and the influence on the flow field is espected to be small. The hot film consists of 24 sensors, which are arranged in flow direction in order to monitor the propagation of disturbances in the boundary layer. As the intent of the hot film measurement is the analysis of the disturbance development in the boundary layer, it is not necessary to calibrate the hot film [5]. The measurements are conducted with a sampling rate of 5 kHz over a time interval of 10 seconds. In addition to the hot film 34 pressure tubes are installed along the chord length in order to gain more information about the pressure distribution. Further measurement equipment like the control units are placed in the nacelle. Unfortunately, measurement parameters like turbulence intensity, wind velocity and resulting rotational speed cannot be chosen freely. Nevertheless, more than 700 measurements, covering a wide range of inflow conditions, have been conducted. Three measurements are chosen for the present analysis. Their inflow parameters, the calculated Reynolds number (Re) and the estimated turbulence level (Tu) are shown in Table 1. In case of $Re = 0.14 \cdot 10^6$ the wind turbine was shut down, but is coasting freeely. For $Re = 1.09 \cdot 10^6$ and $Re = 2.62 \cdot 10^6$ the wind turbine was working with medium or nearly maximum rotational speed, respectively. The turbulence level was calculated based on the incident flow, i.e., the resulting velocity of the wind velocity and the rotational velocity.

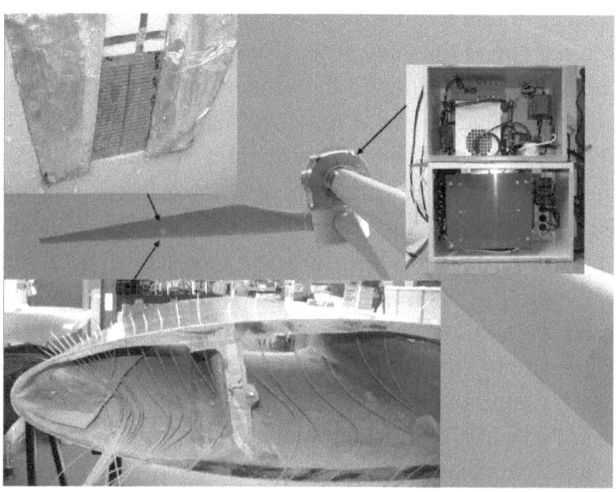

Fig. 1 Measurement set-up including the hot film and the pressure tubes

Table 1 Measurements chosen for the present analysis

Re [–]	v_{wind} [m/s]	v_{rot} [RPM]	Tu [–]
$0.14 \cdot 10^6$	3.24	0.33	0.23
$1.09 \cdot 10^6$	3.02	20.86	0.03
$2.62 \cdot 10^6$	12.57	45.35	0.02

4 Results

In this section the results of the preliminary analysis of the measurements enlisted in Table 1 are discussed. The analysis will concentrate on the hot film measurements. In Fig. 2 (left) the band filtered time signals of all working channels of the hot film are shown for the three different Re numbers. The time signals differ strongly for varying inflow conditions. In case of shut down ($Re = 0.14 \cdot 10^6$) all channels show a rather smooth time signal with only small disturbances. In case of $Re = 1.09 \cdot 10^6$ the first channels show a similar behavior, but the disturbances are obviously amplified with increasing chord length. For $Re = 2.62 \cdot 10^6$ all channels show a time signal with strong disturbances. Further measurements with similar Re numbers show comparable results.

The right graphs in Fig. 2 depict the energy spectra of four hot film channels, which are numbered in flow direction. The energy spectra are calculated from the time signal via fast Fourier transformation (FFT). As expected, the decrease of the energy density as a function of the frequency follows a potential law, similar to the -5/3 law of Kolmogorov. Similar to the time signals the energy spectra illustrate a strong dependency on the inflow conditions. For increasing Re the levels of the energy spectra increase for all channels. In addition to that the graphs in Fig. 2 show strong maxima at 1000 Hz. Up to now these maxima have been categorized as noise and they have been eliminated in the time signals discussed above. Only in case of the medium rotational speed the energy spectra of the four hot film channels shows a characteristic behavior depending on the sensor position at the blade. The energy spectra for channel 5, 12 and 16 are at low level which is similar to the behavior of the measurements at shut down. The energy spectra of channel 20 has a higher level and is comparable to that of the measurement at $Re = 2.62 \cdot 10^6$. Furthermore, a small maximum can be observed for channel 12 and 16 at 300 Hz. This and the behavior of the time series is regarded as a hint for the existence of amplified disturbances in the boundary layer and transition after the TS scenario. Taking all measurement results into account we assume that there is a laminar boundary at $Re = 0.14 \cdot 10^6$; at $Re = 1.09 \cdot 10^6$ the boundary layer seems to be in a transitional state and at $Re = 2.62 \cdot 10^6$ a fully turbulent boundary layer exists [3, 4, 5, 7].

In addition, CFD calculations [8] show that only for $Re = 1.09 \cdot 10^6$ and $Re = 2.76 \cdot 10^6$ amplified disturbances exist in the region of the hot film and that the frequency range of the most strongly amplified disturbances coincide with those of

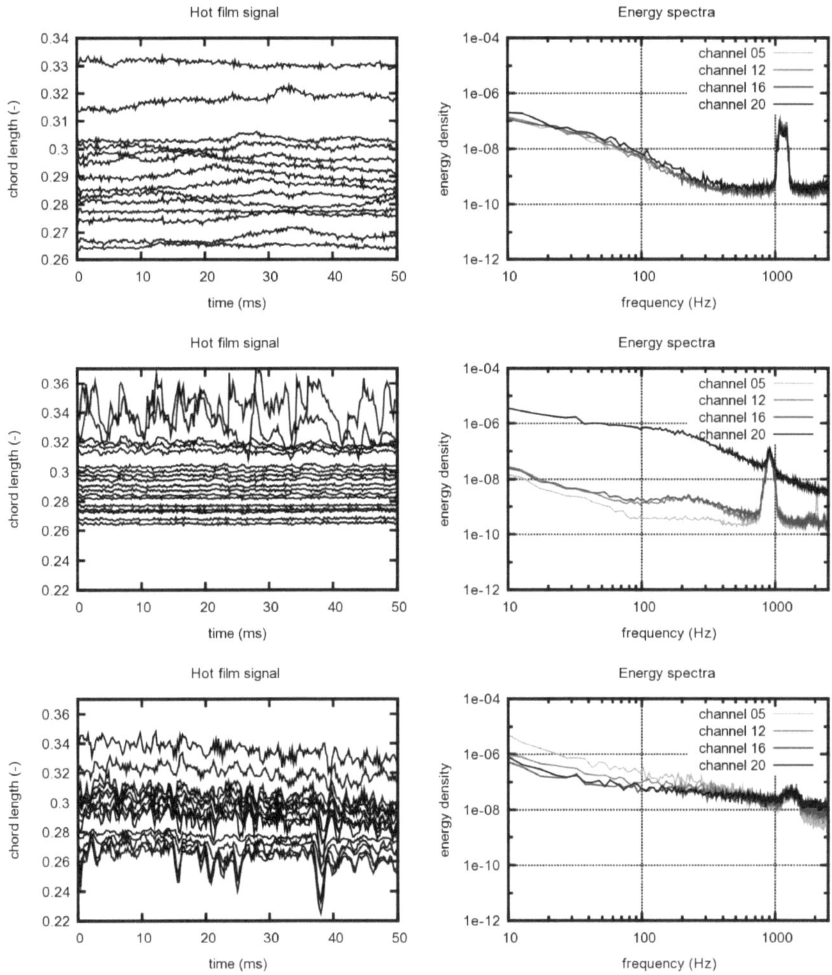

Fig. 2 Time series and energy spectra for hot film measurements at Re = $0.14 \cdot 10^6$ (above), Re = $1.09 \cdot 10^6$ (middle), and Re = $2.62 \cdot 10^6$ (bottom)

the measurements. Thus the results of the CFD calculations support the assumption that in case of a medium rational speed transition occurs after the TS scenario.

5 Conclusions

Measurement equipment, a hot film and pressure tubes, are successfully installed at a wind turbine in order to investigate the boundary layer. During the measurements the turbine is working under real conditions in the atmosphere. The measurements cover a wide range of possible inflow conditions. The pressure measurements show reasonable results and will be presented due to the limited space elsewhere. A preliminary analysis of the hot film measurements suggests that it is possible to gain information about the state of the boundary layer, i.e., whether it is laminar or turbulent. The energy spectra of the measurements with medium rotational speed show for channels near the leading edge a characteristic maximum and an increasing level of the energy spectra along the chord length. In this case a transition process according to the TS scenario is expected, if the behavior of all energy spectra and time series of the measurement signals are taken into account. This assumption has to be verified by comparison with further measurements.

Acknowledgements. We thank the Enercon GmbH for the possibility of installing the measurement equipment at the wind turbine and for the great amount of help during the installation. The project was funded by the Zukunftsprogramm Wirtschaft, which is financed by the European Union, the European Regional Development Fund (ERDF), the Federal Republic of Germany and the Land of Schleswig-Holstein under the Funding Contract No. 122-08-010.

References

1. Kachanov, Y.S.: Physical mechanisms of laminar-boundary-layer transition. Annu. Rev. Fluid Mech. 26, 411–482 (1994)
2. Mack, L.M.: Transition and laminar instability. JPL Publication 77-15. Jet Propulsion Laboratory, California Institute of Technology, Pasadena, California, USA (1977)
3. Madsen, H.A., et al.: The DAN-AERO MW experiments. 48 th AIAA Aerospace Sciences Meeting, Orlando, Florida, January 4-7 (2010)
4. Peltzer, I.: Comparative in-flight and wind tunnel investigation of the development of natural and controlled disturbances in the laminar boundary layer of an airfoil. Exp. Fluids 44, 191–972 (2008)
5. Peltzer, I., Nitsche, W.: Low and high-speed flight experiments on transition detection. Journal of Aircraft 45(6), 1937–1944 (2008)
6. Reed, H.L., Saric, W.S., Arnal, D.: Linear stability theory applied to boundary layers. Annu. Rev. Fluid Mech. 28, 398–428 (1996)
7. Seitz, A.: Freiflug-Experimente zum Übergang laminar-turbulent in einer Tragflügelgrenzschicht. Dissertation, Technical University Carolo-Wilhelmina Braunschweig, Germany (2006)
8. Stock, H.W., Degenhart, E.: A simplified e^N method for transition prediction in two-dimensional, incompressible boundary layers. Z. Flugwiss. Weltraumforsch. 13, 16–30 (1989)
9. White, F.M.: Viscous fluid flow, 3rd edn., pp. 374–379. McGraw-Hill, New York (2006)

Author Index

MIX
Papier aus verantwortungsvollen Quellen
Paper from responsible sources
FSC® C105338

If you have any concerns about our products,
you can contact us on
ProductSafety@springernature.com

In case Publisher is established outside the EU,
the EU authorized representative is:
Springer Nature Customer Service Center GmbH
Europaplatz 3, 69115 Heidelberg, Germany

Printed by Libri Plureos GmbH
in Hamburg, Germany